Capillary Gas
Chromatography

SEPARATION SCIENCE SERIES

Editors: Raymond P.W. Scott, Colin Simpson and Elena D. Katz

Quantitative Analysis using
Chromatographic Techniques

Edited by **Elena D. Katz**

The Analysis of Drugs of Abuse

Edited by **Terry A. Gough**

Liquid Chromatography Column Theory

by **Raymond P.W. Scott**

Silica Gel and Bonded Phases
Their Production, Properties and Use in LC

by **Raymond P.W. Scott**

Capillary Gas Chromatography

by **David W. Grant**

£45.00

LANCHESTER LIBRARY, Coventry University
Gosford Street, Coventry CV1 5DD Telephone 024 7688 7555

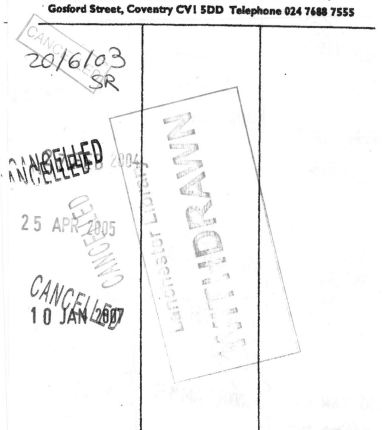

This book is due to be returned not later than the date and time stamped above. Fines are charged on overdue books

Capillary Gas Chromatography

David W. Grant

JOHN WILEY & SONS
Chichester · New York · Brisbane · Toronto · Singapore

Copyright © 1996 by John Wiley & Sons Ltd,
Baffins Lane, Chichester,
West Sussex PO19 1UD, England

Telephone: National 01243 779777
International (+44) 1243 779777

All rights reserved.

No part of this book may be reproduced by any means, or transmitted, or translated into a machine language without the written permission of the publisher.

Other Wiley Editorial Offices

John Wiley & Sons, Inc., 605 Third Avenue,
New York, NY 10158-0012, USA

Jacaranda Wiley Ltd, 33 Park Road, Milton,
Queensland 4064, Australia

John Wiley & Sons (Canada) Ltd, 22 Worcester Road,
Rexdale, Ontario M9W 1L1, Canada

John Wiley & Sons (SEA) Pte Ltd, 37 Jalan Pemimpin #05-04
Block B, Union Industrial Building, Singapore 2057

Library of Congress Cataloging-in-Publication Data

Grant, David W. (David Walter)
    Capillary gas chromatography / David W. Grant.
        p.   cm. — (Separation sciences series)
    Includes bibliographical references and index.
    ISBN 0-471-95377-6 (alk. paper)
    1. Gas chromatography. I. Title. II. Series.
QD79.C45G73   1995
543'.0896—dc20
                                            95-10953
                                                CIP

British Library Cataloguing in Publication Data

A catalogue record for this book is available from the British Library

ISBN 0 471 95377 6

Typeset in 10/12pt Times by Mathematical Composition Setters Ltd, Salisbury, Wilts.
Printed and bound in Great Britain by Bookcraft (Bath) Ltd
This book is printed on acid-free paper responsibly manufacture from sustainable forestation, for which at least two trees are planted for each one used for paper production.

Contents

	Preface	vii
1	General Introduction to Capillary GC	1
2	Theory of Open Tubular Columns	16
3	Capillary Instrumentation	52
4	The Open Tubular Column	107
5	Porous Layer Open Tubular Columns	160
6	Sample Introduction	177
7	Sample Preparation	211
8	Analysis and Optimization	235
9	Multidimensional Capillary GC and Column Switching	267
	Index	289

Preface

The 1958 International Symposium on Gas Chromatography in Amsterdam was the first major occasion at which Marcel Golay presented his paper on the theory of open tubular columns. The present author is one of a privileged group who was fortunate enough to have attended this meeting, and to have experienced first hand the excitement of seeing this powerful separation technique in its embryonic stage. Although many years have now elapsed since this event, it still remains crystal clear in one's memory, particularly the dramatic conclusion in which Golay showed two examples of the new technique's potential. These were the separation of six C6 hydrocarbons in less than nine minutes, and of C8 aromatic hydrocarbons including an almost complete resolution of the m- and p-xylenes. These separations had never been equalled previously on packed columns and certainly did more to impress the audience then the preceding theoretical discussion. I am sure that most of the delegates to this symposium were impatient to return to their laboratories to see if they could duplicate the separations and to apply the technique to their own problems. The author was certainly one of those smitten!

The phenomenal success of capillary GC is now a matter of record. It is a powerful, sophisticated method of analysis employed throughout the chemical industry in all its variety. Nevertheless the diversity of the technique, and the continuing development of new column technology and instrumentation require that any treatise on the subject be continuously updated. This partly is the reason for the preparation of this book, but the author also would hope that some of the early excitement that he experienced could be imparted to a younger generation of chemists whatever their discipline.

The author makes no apology for entitling this book *Capillary Gas Chromatography* although the majority of the columns described herein are more correctly referred to as 'open tubular columns'. Nomenclature has evolved slowly over the past forty years and, as in most branches of science, tends to establish itself by common usage, even if the terminology is often contradictory. For instance, the description of certain types of column as 'wide' or 'mega' bore capillary columns is just one example. With regard to the use of symbols generally, the recent recommendations of the International Union of Pure and Applied Chemistry have been applied as far as possible.

The author would like to thank the many friends and colleagues for their assistance in the preparation of the material, and also the willing support provided by various commercial companies, particularly ex-colleagues at Chrompack International.

Lastly, but above all I would like to thank my wife for her enduring support and encouragement, without which this book would not have been written.

David W. Grant
Chesterfield, Derbyshire
January 1995

1

General Introduction to Capillary GC

1.1 THE CHROMATOGRAPHIC PRINCIPLE

Modern chromatography consists of a range of separation techniques, many of which seem to bear little resemblance to each other. Techniques such as size exclusion, ion exchange, and capillary gas chromatography have diverse fields of application, and require very different facilities to operate them, but they do have a common principle in that they are all based on different rates of migration of sample components through a static retentive medium. The static phase may have either a planar configuration, as in paper and thin layer chromatography, or confined by a tube, as in column chromatography.

Figure 1.1 shows the basic range of column techniques and some of the descriptive nomenclature employed.

In column chromatography the sample is transported through the static medium in a carrier fluid called the **mobile phase**. In **gas chromatography (GC)** the mobile phase is commonly referred to as the **carrier gas**. The retentive part of the static medium is called the **stationary phase** and can consist of a liquid or solid material effectively distributed inside the column to maximize contact with the mobile phase.

In **partition chromatography** the stationary phase is conventionally a liquid, and chromatographic techniques were originally identified according to the physical state of the mobile and stationary phase respectively, as in liquid–liquid partition chromatography (LLPC) and gas–liquid partition chromatography (GLPC). Nowadays it is often difficult to define the physical state of the stationary phase or the precise mechanisms involved and so this early terminology has been replaced by the less ambiguous terms **liquid chromatography (LC or HPLC)** and **gas chromatography (GC)**. Any presumption of the separation mechanism, i.e., whether it is based on adsorption, partition, etc., is thereby avoided. The stationary phase in gas chromatography for instance is often immobilized by a polymerization process

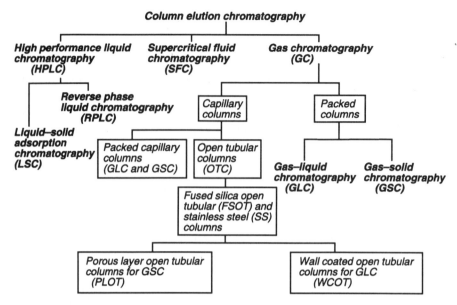

Figure 1.1 Column chromatographic techniques and nomenclature

and may have the appearance of an amorphous solid, but nevertheless it behaves as a conventional liquid.

Whatever is the retention mechanism, an important feature of most of the modern techniques is that the solute distribution is linear, i.e., that the concentration of each component in the stationary phase is proportional to its concentration in the mobile phase so that the ratio of the two concentrations is constant under equilibrium conditions. Symmetry of peak shape is largely dependent in this linearity and this in turn affects the separation and accuracy of quantitative analysis. In **adsorption chromatography** linearity means that the adsorption isotherm is linear, and in **partition chromatography**, it means that the partition coefficient is constant.

Another important practical requirement of modern chromatographic methods is frequent deviation from ideal behaviour. For instance, the ability of GC to separate compounds with similar vapour pressure characteristics depends on deviations from Raoult's law whereby selective interactions with the stationary phase are exploited.

1.2 ORIGINS OF CHROMATOGRAPHY

Modern chromatography has developed almost entirely during the past fifty years and it is now widely employed for both qualitative and quantitative analysis in virtually every chemical field. Of all of the chromatographic techniques, capillary gas chromatography is the most powerful with regard to its ability to separate very complex mixtures.

The term 'chromatography' was first applied by the botanist, M. S. Tswett in 1903 [1] to describe the technique he devised for the separation of plant pigments. These compounds produced separate coloured zones on columns packed with adsorbent materials when plant extracts were percolated through them. Tswett is now justifiably regarded as the inventor of chromatography although of course his early techniques bear little resemblance to modern methods. Nevertheless, the physical process involving a continuous equilibration between two phases is the common factor and so the term 'chromatography' is now accepted for all methods based on the same principle.

Zswett published numerous papers in the period 1903–1910, but, as is so often the case, his results were heavily criticized by his contemporaries. Consequently, he did not receive the recognition he deserved during his lifetime and the technique remained dormant for some time. A contributing factor to this could also have been the relative inaccessibility of the journals in which the work was published, coupled with the translational difficulties. Even today it is difficult to find copies of these first publications although there is now an English translation of his original paper [2]. Biographical reviews on the life of Tswett have been published by Dhere [3], Robinson [4], and, more recently a compilation by Berezkin [5] has been published.

Much of the credit for reviving the chromatographic technique must be attributed to the American scientist L. S. Palmer who used the technique extensively for the separation of natural products, and documented it in his monograph published in 1922 [6]. This book later motivated other researchers working in the field of natural products, notable Kuhn and his assistants, Winterstein, Lederer, and Brockman. Kuhn's group were studying the carotenoids in particular and were successful in separating many of these important isomers [7–12].

The chromatographic techniques employed at the time were all based on adsorption in large diameter columns packed with a variety of adsorbents such as alumina and calcium carbonate, using strong organic solvents as eluents. They were early examples of **elution** chromatography carried out on **liquid–solid adsorption** columns, with the primary purpose of recovering sufficient quantities for identification by other means.

The period 1940–50 saw major developments in chromatography and these laid the foundations to the modern technique. The most significant event of this period was undoubtedly the introduction of **partition chromatography** by Martin and Synge in 1941 [13,14] who were working on ways of improving the analysis of proteins. They found that a column packed with silica gel, heavily impregnated with water, could be operated as a partition system by percolating chloroform through it. The water was retained strongly by the silica gel which effectively immobilized it as a stationary phase, while the immiscible chloroform functioned as a mobile phase. Components thereby separated according to the differences in their distribution coefficients, and were detected by the use of colour indicators.

This type of chromatography was a major breakthrough for a variety of reasons but the two most important of these are as follows.

1. Partition coefficients are constant and independent of solute concentrations under dilute solution conditions. This causes solute zones to be symmetrically distributed inside the column thereby allowing a complete quantitative separation to be achieved between neighbouring zones.
2. The plate theory as applied by Martin and Synge to linear chromatography showed that the zone concentration profile has a Gaussian character thus providing with a quantitative way of describing the quality of columns in terms of column efficiency or theoretical plate number.

The importance of this work was recognized in 1952 by the award of the Nobel prize in chemistry for these two scientists.

In retrospect, it is significant that in their publications Martin and Synge speculated on the possibility of using a gas or vapour as a mobile phase in partition chromatography. In fact gas chromatography using adsorption columns was beginning to emerge in this period, mainly through the work of Tiselius and Claesson [15–18] who defined the three principle techniques of **frontal analysis, displacement,** and **elution chromatography**. Further progress in the use of displacement gas chromatography on charcoal columns was made at this time by Phillips (1949) in connection with gas kinetics studies [19], while Cremer and her group was developing gas elution techniques using both charcoal and silica gel columns [20–22]. The now familiar **thermal conductivity detector** (**TCD**) was first employed at that time.

It is, perhaps, rather surprising that the early suggestion of using a gaseous mobile phase in **partition** columns was not taken up during this period, although a possible explanation could have been the nature of the projects under study. Gas **adsorption** chromatography developed almost incidentally as a way of analysing gas mixtures resulting from unconnected research studies. This is also probably true in relation to much of the early progress in chromatography. In later years more and more interest was directed to developing the techniques themselves, particularly as the potential for expanding the scope to a much wider range of compounds was realized.

In 1952, Martin, in collaboration with his colleague James published the first description of a **gas–liquid partition** system, and described its application to the separation of volatile fatty acids, which even today is recognized as a difficult application [23]. The all-glass apparatus used an ingenious autotitration system to detect acidic and basic compounds. The enormous industrial potential of this new technique was quickly realized, particularly within the petrochemical industry, and early pioneers such as Ray [24], and Bradford, Harvey and Chalkley [25] described working systems which they established in their application laboratories. The period 1952–60 produced a rapid growth of interest in the new technique and new research results were soon to be presented and discussed in a number of symposia that were organized at the time. The newly formed Institute of Petroleum Gas Chromatography Discussion Group was responsible for much of this early activity in the UK.

A typical gas chromatographic column of this early period consisted of a simple glass or metal tube of up to several metres in length packed with a mixture of a diatomaceous earth impregnated with 5–20% of a high boiling organic solvent as the stationary phase. This would be heated by a thermostatically controlled oven and supplied with a controlled flow of an inert carrier gas, usually nitrogen. Thermal conductivity detectors, then known as katharometers were normally employed for detection.

1.3 DEVELOPMENT OF CAPILLARY GAS CHROMATOGRAPHY

Golay (1957) first suggested the use of coated open tubular columns for gas chromatography as a result of a mathematical study. The results were initially presented at the First Gas Chromatography Symposium of the Instrument Society of America in East Lancing, Michigan [26]. The following year, a more detailed presentation was given at the Second International Symposium on Gas Chromatography in Amsterdam [27].

We can summarize the main advantages of an open tubular system as follows.

1. Rates of diffusion in gases are some four to five orders of magnitude faster than in liquids. This means that with a gaseous mobile phase there does not need to be such intimate contact between the two phases in order to achieve fast equilibration. Hence columns need not be packed conventionally provided that the stationary phase is present as a thin film on the inner surface of the column, and that the column diameter is sufficiently small. Thus column diameters of 0.3–0.4 mm will still give fast equilibration and consequently high column efficiencies.
2. In an open tubular column the pressure drop per unit length is several orders of magnitude lower than in packed columns. This means that very much longer open tubular columns can be used without excessive pressure drops, thereby producing much better separations.
3. The absence of the packing material eliminates one of the causes of zone dispersion inside the column (see Chapter 2).
4. The absence of a so-called inert support removes one of the major causes of peak tailing in columns. Support materials are usually based on the use

of diatomaceous earths and are silicacious in character. This means that they have active sites due to the presence of hydroxyl groups and metal impurities which can interact with sample components and cause adsorption effects and hence peak tailing.

5. There is usually a much smaller quantity of stationary phase in an open tubular column than in a packed column. This reduces the retentive properties of the column thereby reducing analysis time.

In spite of the excitement that these new developments generated at the time, there were considerable difficulties to be overcome before they could become viable. GC equipment had a fairly basic design which was completely inadequate for use with capillary columns because of their much smaller volume and minute sample requirements. The principal difficulties were the low sensitivities and relatively large cell volumes of the detectors used with packed columns. It is perhaps rather fortuitous that the **flame ionization detector, (FID)** [28] had also been introduced at the 1958 Amsterdam symposium. This detector not only had the requisite small volume for capillary operation, but it also possessed a very high sensitivity. This detector was to become the most ubiquitous detector in capillary GC and is nowadays regarded as the 'standard' detector in capillary instrumentation. McWilliam and Dewar are often credited with the invention of the FID, but the device was in fact first described by Pretorius and his group [29] a few months earlier. The FID was undoubtedly inspired by an earlier flame detector, described by Scott (1955), which was based on measuring variations in the heat of a hydrogen flame burning at the column exit as components eluted [30].

In the following two years, the potentialities of the new columns were investigated by such pioneers as Condon [31], Desty [32–35], Kaiser and Struppe [36], Lipsky and coworkers [37], and Scott [38–40].

During the years that have elapsed since its introduction capillary gas chromatography has become one of the most powerful separation techniques available to the analytical chemist. In fact it would be difficult to find any activity in the general field of organic chemistry which does not involve this remarkable method in some form or other. Of particular importance today are those disciplines involving petrochemicals, pharmaceutical, environmental monitoring, food and beverage chemistry, and even clinical studies. Figure 1.2, for instance, is an example of a modern capillary chromatogram showing the separation of flavour compounds.

Of course, as is the case with most separation methods, capillary GC is essentially a practical technique, relying heavily on high quality instrumentation, especially in relation to efficient sample introduction, highly sensitive detectors, and powerful data systems. Nevertheless, many of Golay's theoretical predictions in relation to the effects of parameters and operating variables have proved to be valid. In spite of their potential attractions early

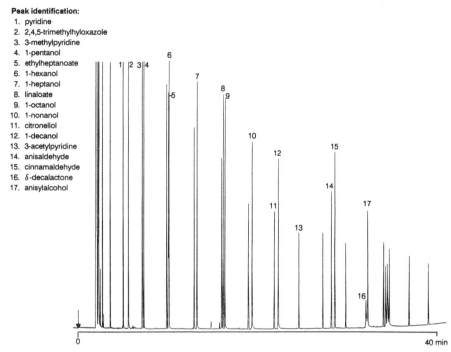

Figure 1.2 Separation of some flavour compounds by capillary GC. 50 m × 0.32 mm diameter fused open tubular (FSOT) column coated with CPwax52CB (a chemically bonded polyethylene glycol); film thickness = 0.2 μm. Temperature programme: 60 to 260 °C at 3 °C min^{-1}; nitrogen carrier gas with split injection and flame ionization detector. Sample size 0.5 μl of solution. (Reproduced by permission of Chrompack International BV)

Peak identification:
1. pyridine
2. 2,4,5-trimethylhyloxazole
3. 3-methylpyridine
4. 1-pentanol
5. ethylheptanoate
6. 1-hexanol
7. 1-heptanol
8. linaloate
9. 1-octanol
10. 1-nonanol
11. citronellol
12. 1-decanol
13. 3-acetylpyridine
14. anisaldehyde
15. cinnamaldehyde
16. δ-decalactone
17. anisylalcohol

open tubular columns had a poor reproducibility and a short lifetime. Patented stainless steel columns were available commercially but these had a limited application, and parameters such as film thickness were generally poorly defined. Desty and coworkers (1960) developed an apparatus for making long lengths of capillary glass tubing [41], and this material soon became a popular option but methods of coating the tube with stationary phase were still unsatisfactory from the viewpoint of reproducibility. Other problems arose as a result of the very small amounts of stationary liquid phase on the capillary wall. The film was often unstable particularly at elevated temperatures. Also, because of the adsorptive nature of the capillary surface, particularly in stainless steel columns, polar compounds often gave severe peak tailing.

The milestones in the development of capillary GC are summarized in Table 1.1.

Table 1.1 Milestones in the development of capillary GC

Year	Event	Author(s)	References
1957–1958	Introduction of open tubular columns	Golay	26,27
1958	Flame ionization detector	McWilliam and Dewar	28
		Harley, Nel and Pretorius	29
1960	Electron capture detector	Lovelock and Lipsky	Chapter 3, ref. 22
1960	Glass capillary columns	Desty, Haresnape and Whyman	41
1962–on	Porous layer open tubular (PLOT)	Various workers	See Chapter 5
1974–on	Chemically bonded (CB) stationary phases	Various workers	See Chapter 4
1979	Flexible fused silica (FS) capillary columns	Dandeneau and Zerenner	42

1.4 BASIC CAPILLARY SYSTEM

A detailed description of the instrumentation associated with modern capillary GC is given in Chapter 3. Figure 1.3 shows the basic layout.

The carrier gas usually consists of pure, oxygen-free helium, and is supplied to the column by a **flow control system**. This is a combination of pressure and flow controllers which ensures that the column and associated hardware, such as the detector and sample introduction system, operate under appropriately controlled flow conditions.

The **injector** is the interface between the operator and the instrument by which the sample is introduced into the column. The different ways of accomplishing this are discussed in Chapter 6. Most injectors are operated at elevated temperatures and so they require controlled heating, typically to a maximum temperature of 360–400 °C. The temperature controlled environment for the column is called the column oven and this must be both precise and accurate since the column temperature influences the retention characteristics in a critical manner. When the column temperature is kept constant throughout the run the operational mode is referred to as **isothermal operation**. Most instruments are equipped with versatile temperature control systems which also

GENERAL INTRODUCTION TO CAPILLARY GC

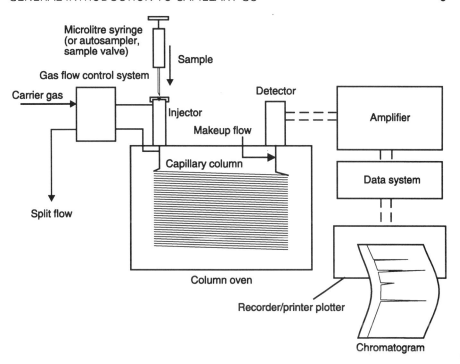

Figure 1.3 Schematic diagram of universal gas chromatograph

enable the oven to be heated in a controlled fashion during the run. This mode is called **programmed temperature operation**. The majority of modern ovens are also designed to allow the use of two channels, namely, two columns which can be used independently with separate injectors and detectors.

The **detectors** are usually mounted on top of the oven and may consist either of similar or different types depending on the application. These are sensitive devices which are connected to the end of the column and respond to the presence of eluted compounds from the column, hence producing an electrical signal which is (normally) proportional to the amount of that compound.

Modern instruments are usually equipped with **data systems** which receive the detector output, store it in digital format, and process the data for the purposes of analysis. Normally, the data system output is connected to a printer plotter which produces a chromatogram, in which each separated component gives a peak, and the results of the analysis. The data system is not essential to the chromatography, however, and results can be calculated from manual measurements made from a simple chart recorder connected directly to a detector amplifier. However, such a procedure would be very time-consuming and is generally not regarded as a viable option in busy quality control laboratories.

1.5 FURTHER NOMENCLATURE AND DEFINITIONS

Much of the nomenclature for chromatography has been updated recently by IUPAC [43] and this book will conform to these recommendations. The tube containing the retentive medium in which the separation is carried out is called the **column**. In normal packed partition columns this medium consists of an inert porous material, usually a granular diatomaceous earth known as the **support**, impregnated with between 3–30% by weight of a liquid **stationary phase**.

Figure 1.4 shows diagrammatical representations of the cross-sections of the three major types of capillary column in current use. In **open tubular columns**, the stationary phase is coated as an even layer over the inner wall of the tube. This phase can be either a liquid, sometimes highly cross-linked to be solid in appearance, or it can be a porous layer of solid adsorbent material such as alumina, or molecular sieve. The latter columns are known as **porous layer open tubular (PLOT)** columns and make a very important contribution to modern capillary GC. Most open tubular columns are now constructed either from fused silica or stainless steel. **Packed capillary columns** have a similar tube geometry to open tubular columns except that they are filled with packing material. The packing is accomplished in a variety of ways, but usually the packing density is rather less than in conventional packed columns thus giving a higher permeability (cf. Table 4.1). Packed capillary columns have not become popular but they are sometimes applied to some specialized applications.

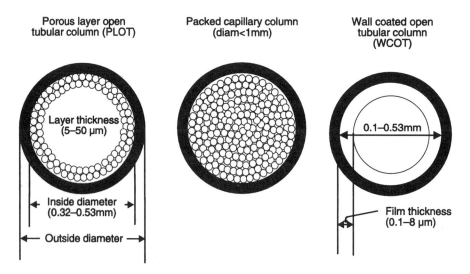

Figure 1.4 Types of capillary column

GENERAL INTRODUCTION TO CAPILLARY GC

One of the most attractive features of the open tubular column is its accurate and well-defined geometry, and modern columns are made to a precise specification in which the geometrical parameters are known accurately. These parameters are length, column diameter, and stationary phase film thickness.

A fundamental property of the open tubular column is the **phase ratio**, β defined as follows:

$$\text{phase ratio}(\beta) = \frac{\text{volume of carrier gas in column}}{\text{volume of stationary phase in column}} \quad (1.1)$$

From the known geometry of the column, the phase ratio can be calculated:

$$\beta = \left(\frac{d_i}{4d_f}\right) \quad (1.2)$$

where d_i = internal column diameter, and d_f = thickness of the stationary phase film in the same units.

The retention properties of components depends critically on this ratio, and also on the affinity of compounds for the stationary phase, as measured by the partition coefficient K, defined as follows:

$$K = \frac{\text{concentration in stationary phase}}{\text{concentration in carrier gas}} \quad (1.3)$$

where the concentrations are measured in terms of mass/volume per cent. In practice it is more useful to convert this to the **retention factor**, k, defined as the ratio of the mass of each component in the stationary phase to its mass in the carrier gas at equilibrium. The retention factor is obtained from the partition coefficient as follows:

$$\text{retention factor } k = \frac{\text{mass of solute in stationary phase}}{\text{mass of solute in carrier gas}}$$

$$= \frac{K}{\beta} \quad (1.4)$$

$$= \frac{\text{time spent in stationary phase}}{\text{time spent in carrier gas}} \quad (1.5)$$

The retention factor can be easily calculated from the measured retention times in the chromatogram, as shown in Figure 1.5. The time interval between sample injection and the peak maximum is the **total retention time** t_R but for the calculation of k this must be corrected for the column **holdup time** t_0 to give the **adjusted retention time** t'_R, which is the actual time spent by the solute in the stationary phase. The holdup time is the retention time of a

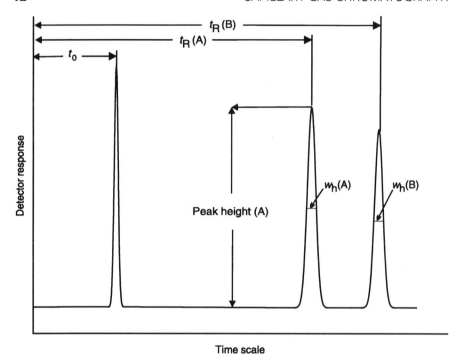

Figure 1.5 Basic chromatographic measurements. Key: t_o = column holdup time; $t_R(A)$ = total retention time for A; $t_R(B)$ = total retention time for B; $w_h(A)$ = peak width at half height for A; $w_h(B)$ = peak width at half height for B

component which travels through the column at the same velocity as the carrier gas, i.e., it is not retarded by the stationary phase and therefore elutes from the column in the shortest possible time. Such a component is referred to as a **non-retained component**. The terms **void** or **dead time** are often used to describe the holdup time but they tend to have rather ambiguous meanings and so they should be avoided in this context.

The adjusted retention time is therefore given by,

$$t'_R = t_R - t_0 \qquad (1.6)$$

hence

$$k = \frac{t'_R}{t_0} \qquad (1.7)$$

The **mean linear velocity**, $\bar{u}$, of the carrier gas is calculated from the column length, L, and the holdup time, namely,

$$\bar{u} = \frac{\text{column length}}{\text{holdup time}} = \frac{L}{t_0} \qquad (1.8)$$

A mean value of gas velocity is necessary in GC because the gas velocity increases along the column as the gas decompresses. In Chapter 2 we shall see how the mean gas velocity affects the column performance, and the relationship between the inlet, exit, and mean values of velocity. The carrier gas velocity is critical from the viewpoint of column performance and it is important for it to be set at an appropriate value. With regard to column holdup time, this for most columns, can be measured by injecting methane, assuming that this hydrocarbon is not retained significantly by the stationary phase.

The plate theory of chromatography was mentioned earlier and will be discussed in some detail in Chapter 2. Columns are characterized by their **theoretical plate number**, N, also called the **column efficiency**, calculated from a chromatogram run under isothermal conditions from the following relationship:

$$N = 5.54 \times \left(\frac{t_R}{w_h}\right)^2 \tag{1.9}$$

where w_h is the width of the peak at half height as illustrated in Figure 1.5.

The compound chosen for this measurement must be reasonably well retained, with a k value greater than four. Typical plate numbers for capillary columns can range from several thousand to several hundred thousand. The column efficiency defines the width of the chromatographic zone inside the column just before it emerges and enters the detector. Columns with high theoretical plate numbers produce very narrow peaks which increases the likelihood of separating all the components of complex mixtures.

In general, column efficiency increases proportionally with the length of the column, i.e. doubling the column length will provide twice the number of theoretical plates. It is often more meaningful to use the **theoretical plate height**, H, calculated by dividing the length of the column by its number of theoretical plates, namely:

$$H = \frac{L}{N} \tag{1.10}$$

Although the theoretical plate number is a measure of column performance, it does not indicate the resolving power of the column. Thus two components having the same partition coefficients will travel along the column at exactly the same velocity and so they will elute unresolved at the same time. In order to assess the ability of the column to resolve any two compounds A and B we must also include the effect of differences in their partition coefficients. The **separation factor**, α, is defined as the ratio of the two retention factors, calculated using the larger of the two values as the numerator, as follows:

$$\text{separation factor } \alpha = \frac{k_B}{k_A} \geq 1 \tag{1.11}$$

For capillary GC to separate the compounds successfully, the separation factor must be greater than unity by an amount depending on the number of theoretical plates available. Open tubular columns, for instance, require smaller separation factors than packed columns to achieve effective separations because of their much higher plate numbers. The quantitative expression of separation is **resolution** R_s calculated from the following expression:

$$R_s = \frac{1.2[t'_R(B) - t'_R(A)]}{[w_h(A) + w_h(B)]} \qquad (1.12)$$

Defining resolution in this way quantifies the amount of peak area overlap between the two peaks. This is obviously important from the point of view of quantitative analysis where mutual interference should be minimal. An acceptable overlap for most practical purposes is less than one per cent of the respective peak areas which is equivalent to a resolution of less than 1.5. A detailed discussion of this topic is given in Chapter 2.

1.6 SUMMARY

Modern capillary GC is the culmination of many years of effort by numerous research workers in universities, industry, and not least by the research and development departments of commercial companies dealing with instruments and consumables. The diverse character of this effort has resulted in a great variety of instruments, techniques, and columns covering an enormous range of applications.

REFERENCES

1. Tswett, M. S. (1903) *Tr. Protok. Varshav. Obshch. Estestvoispyt. Otd. Biol.* (Transactions of the Warsaw Naturalist Society) **14**.
2. Hesse, G. and Weil, H. (1954) *Michael Tswett's First Paper on Chromatography*, M. Woelm, Eschege.
3. Dhere, C. (1943) *Candollea*, **10**, 23.
4. Robinson, T. (1960) *Chymia*, **6**, 146.
5. Berezkin, V. G. (Compiler) (1990) *Chromatographic Adsorption Analysis: Selected Works of Mikhail Semenovich Tswett*, Ellis Horwood Series in Analytical Chemistry, Chichester.
6. Palmer. L. S. (1992) *Carotenoids and Related Pigments: The Chromolipids*. Am. Chem. Soc. Monograph Series, Chemical Catalog Co., New York.
7. Kuhn, R. *et al.* (1931) *Hoppe-Seyler's Z. Physiol. Chem.*, **197**, 141.
8. Kuhn, R. and Lederer, E. (1931) *Hoppe-Seyler's Z. Physiol. Chem.*, **200**, 246.
9. Kuhn, R. and Brockman, H. (1931) *Ber.*, **64**, 1859.
10. Kuhn, R. and Lederer, E. (1931) *Hoppe-Seyler's Z. Physiol. Chem.*, **200**, 108.
11. Kuhn, R. and Brockman, H. (1932) *Hoppe-Seyler's Z. Physiol. Chem.*, **206**, 41.

12. Kuhn, R. and Brockman, H. (1933) *Ber.*, **66**, 407.
13. Martin, A. J. P. and Synge, R. L. M. (1941) *Biochem. J,.* **35**, 91.
14. Martin, A. J. P. and Synge, R. L. M. (1941) *Biochem. J.*, **35**, 1358.
15. Tiselius, A. (1943) *Kolloid Z.*, **105**, 101.
16. Claesson, S. (1948) *Ann. N.Y. Acad. Sci.*, **49** 183.
17. Claesson, S. (1946) *Ark. Kem. Mineral. Geol.*, **23A** (1).
18. Tiselius, A. and Claesson A. (1942) *Ark. Kem. Mineral. Geol.*, **15B**, (18).
19. Phillips, C. S. G. (1949) *Disc. Faraday. Soc.*, **7**, 241.
20. Cremer, E. (1976) *Chromatographia*, **9**, 364.
21. Cremer, E. and Prior, F. (1949) *Osterr. Chem. Ztg.*, **50**, 161.
22. Cremer, E. and Prior, F. (1951) *Z. Electrochem.*, **55**, 66.
23. James, A. T. and Martin, A. J. P. (1952) *Biochem. J. (London)*, **50**, 679.
24. Ray, N. H. (1954) *J. Appl. Chem. (London)*, **4**, 21.
25. Bradford, B. H., Harvey, D.and Chalkley, D.E. (1955) *J. Inst. Petroleum*, **41**, 80.
26. Golay, M. J. E. (1958) in *Gas Chromatography 1957 (Lansing Symposium)*, (eds V. J. Coates *et al.*), Academic Press, New York, pp. 1–13.
27. Golay, M. J. E. (1958) in *Gas Chromatography 1958 (Amsterdam Symposium)*, (ed. D. H. Desty), Butterworths, London, pp. 139–143.
28. McWilliam, I. G. and Dewar, R. A. (1958) *Nature (London)*, **181**, 760.
29. Harley, J., Nel, W. and Pretorius, V. (1958) *Nature (London)*, **181**, 117.
30. Scott, R. P. W. (1955) *Nature (London)*, **176**, 793.
31. Condon, R. D. (1959) *Anal. Chem.*, **31**, 1717.
32. Desty, D. H. (1959) in *Gas Chromatographie 1958*, (ed. H. P. Angele), Akademie Verlag, Berlin (Ost), pp. 176–184.
33. Desty, D. H. and Goldup, A. (1960) in *Gas Chromatography 1960*, (ed. R. P. W. Scott), Butterworths, London, pp. 162–183.
34. Desty, D. H., Goldup, A. and Swanton, W. T. (1962) in *Gas Chromatography*, (eds N. Brenner *et al.*), Academic Press, New York, pp. 105–135.
35. Desty, D. H., Goldup, A. and Whyman, B. H. F. (1959) *J. Inst. Petrol.*, **45**, 287.
36. Kaiser, R. E. and Struppe, H.G. (1959) in *Gas Chromatographie 1959*, (ed. R. E. Kaiser and H. G. Struppe), Akademie Verlag, Berlin (Ost), pp. 177–194.
37. Lipsky, S. R., Lovelock, J. E. and Landowne, R. A. (1959) *J. Am. Chem. Soc.*, **81**, 1010.
38. Scott, R. P. W. (1959) *Nature*, **183**, 1753.
39. Scott, R. P. W. (1961) *J. Inst. Petrol.*, **47**, 284.
40. Scott, R. P. W. and Hazeldean, G. S. F. (1969) in *Gas Chromatography 1960*, (ed. R. P. W. Scott), Butterworths, London, pp. 144–161.
41. Desty, D. H, Haresnape, J. N. and Whyman, B. H. F. (1960) *Anal. Chem.*, **32**, 302.
42. Dandeneau, R. D. and Zerenner, E. H. (1979) *J. High Resolut. Chromatogr.*, **2**, 351.
43. Ettre, L. S. (1993) *Nomenclature for Chromatography*, IUPAC recommendations 1993, *Pure Appl. Chem.*, **65**, 819.

2

Theory of Open Tubular Columns

2.1 INTRODUCTION

Most of the chromatographic techniques which are used today have evolved in a fairly haphazard and empirical manner, based on a need to develop effective ways of separating compounds which are of direct interest to the user. In contrast, modern capillary GC has developed largely as a consequence of Golay's original theoretical concepts. We shall see that much of this theory is applicable to the modern column.

The author is well aware of the resistance to the use of theory shown by many practitioners of the subject, in what could be regarded as an essentially practical technique. Nevertheless, experience has shown that an appreciation of the theory can facilitate the selection of appropriate column dimensions and operating conditions. This chapter has been prepared from the point of view of the working analyst to provide a sound understanding of the quantitative relationships between the operating variables and resulting column performance.

2.2 EFFECTS OF GAS COMPRESSIBILITY

One of the most significant differences between GC and HPLC is the compressible nature of the gaseous mobile phase. According to Boyle's law the gas velocity increases along the column in inverse proportion to the decrease in pressure, so that the ratio of the gas velocity at the column exit to its velocity at the inlet is equal to the pressure drop ratio, i.e.,

$$\frac{u_0}{u_i} = \frac{p_i}{p_0} \tag{2.1}$$

where u_0 = exit carrier gas velocity; u_i = inlet velocity; p_i = inlet pressure; p_0 = exit pressure (normally atmospheric). Consider a section with a thickness

THEORY OF OPEN TUBULAR COLUMNS

dx at distance x from the column inlet. If the pressure drop across this section is dp producing a gas velocity of u_x cm s^{-1} then u_x is related to dp by,

$$u_x = -\frac{B_0}{\eta}\frac{dp}{dx} \qquad (2.2)$$

where η is the gas viscosity and B_0 is the **column permeability**, a constant which measures its resistance to gas flow. Open tubular columns have intrinsically high permeabilities; this is one of their major advantages over packed columns since it allows very much longer columns to be employed for the same level of pressure drop, thereby generating much higher theoretical plate numbers.

According to Boyle's law

$$p_x u_x = p_0 u_0 \qquad (2.3)$$

where p_x and u_x are the pressure and velocity of the gas at point x.

$$u_x = u_0\left(\frac{p_0}{p_x}\right) \qquad (2.4)$$

By substituting (2.2) into (2.4)

$$\left(\frac{p_0}{p_x}\right)u_0 = -\frac{B_0}{\eta}\frac{dp}{dx} \qquad (2.5)$$

namely

$$\eta p_0 u_0 \int_0^x dx = -B_0 \int_{p_i}^{p_x} p \cdot dp \qquad (2.6)$$

and integrating between 0 and x we obtain

$$\eta p_0 u_0 x = \frac{B_0}{2}(p_i^2 - p_x^2) \qquad (2.7)$$

When $x = L$ and $p_x = p_0$,

$$\eta p_0 u_0 L = \frac{B_0}{2}(p_i^2 - p_0^2) \qquad (2.8)$$

dividing (2.7) by (2.8)

$$\frac{x}{L} = \frac{p_i^2 - p_x^2}{p_i^2 - p_0^2} \qquad (2.9)$$

$$= \frac{P^2 - \left(\dfrac{p_x}{p_0}\right)^2}{P^2 - 1} \tag{2.10}$$

where $P = p_i/p_0$; hence, from (2.4)

$$\frac{u_x}{u_0} = \frac{1}{[P^2 - (x/L)(P^2 - 1)]^{0.5}} \tag{2.11}$$

Figure 2.1 shows plots of u_x/u_0 versus x/L for various pressure drops.

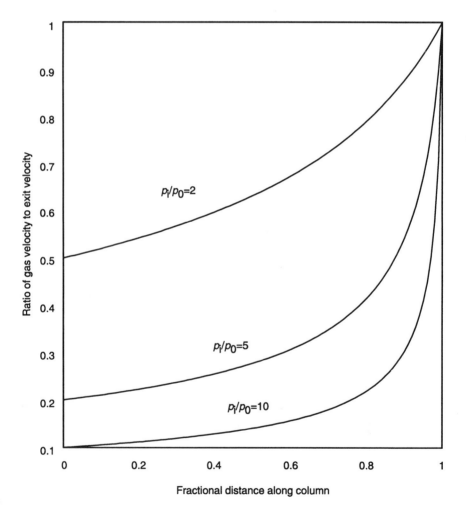

Figure 2.1 Carrier gas velocity profiles for various pressure drops

THEORY OF OPEN TUBULAR COLUMNS 19

For low pressure drops the increase in gas velocity is almost linear with column length, and in the hypothetical situation of zero pressure drop then the gas velocity would be constant throughout the column.

For high pressure drops we see that there is a rapid increase in the gas velocity near the column exit while most of the column is operating well below the exit velocity. In fact the inlet velocity then becomes far more representative of the **average** velocity in the column than the exit velocity. This can be problematical for some applications, particularly for the measurement of absolute retention data or if the column were to be operated according to the **exit** flow conditions. This problem is avoided by using retention data based on **relative** measurements and by operating the column at a selected **mean carrier gas velocity**, $\bar{u}$.

2.2.1 Mean (Average) Gas Velocity $\bar{u}$

It is clear from Figure 2.1 that since the gas velocity does not increase linearly along the column then the mean gas velocity cannot be defined as the average of the inlet and exit velocities. The mean gas velocity, $\bar{u}$, is defined as the ratio of the column length to the holdup time, namely, $\bar{u} = L/t_0$, and the relationship between this mean value and the exit velocities u_0 is derived as follows.

The column holdup time t_0 is given by:

$$t_0 = \int_0^L \left(\frac{dx}{u_x}\right) \tag{2.12}$$

$$= \int_0^L \frac{p_x}{p_0 u_0} \cdot dx$$

and substituting the value of dx derived from (2.2) gives

$$t_0 = -\int_0^L \frac{B_0 p_x^2}{p_0^2 u_0^2 \eta} \cdot dp \tag{2.13}$$

$$= \frac{B_0(p_i^3 - p_0^3)}{3\eta p_0^2 u_0^2} \tag{2.14}$$

but $\bar{u} = L/t_0$; therefore

$$\bar{u} = \frac{3\eta L p_0^2 u_0^2}{B_0(p_i^3 - p_0^3)} \tag{2.15}$$

and by substitution from (2.8)

$$\bar{u} = u_0 \frac{3(p_i^2 - p_0^2)p_0}{2(p_i^3 - p_0^3)} \tag{2.16}$$

$$= u_0 \frac{3[(p_i/p_0)^2 - 1]}{2[(p_i/p_0)^3 - 1]} \tag{2.17}$$

i.e. $\bar{u} = u_0 j$ where

$$j = \frac{3[(p_i/p_0)^2 - 1]}{2[(p_i/p_0)^3 - 1]} \tag{2.18}$$

$$= \frac{3(P^2 - 1)}{2(P^3 - 1)} \tag{2.19}$$

The factor j, known as the **Martin compressibility correction factor**, relates the exit gas velocity to the mean value and is used to correct measured data to a hypothetical zero pressure drop where $P = 1$. It is mostly used to determine **absolute** retention data using packed columns where the measurement of column **exit** flow rates is the normal practice. Because of gas compressibility, measured retention times do not change in inverse proportion to the exit flow rate and so the j factor must be used to correct the measurements if absolute data is required. This correction is rarely necessary in capillary GC because the **average** gas velocity is simple to measure in practice and optimum operating conditions are normally based on this setting. Also, **relative** retention data is not only easier to measure but is usually more accurate than absolute data for identification purposes.

2.3 THE SEPARATION PROCESS

Two types of theoretical treatment have evolved during the past fifty years which are of practical importance, namely, the **plate theory** and the **kinetic**, or **rate** theory. The plate theory provides us with our standard method for the measurement of column quality via the **theoretical plate number**, N. The rate theory on the other hand describes the various diffusion processes which are responsible for band broadening. Before we discuss these theories in detail let us first examine the character of the overall separation process.

Assume that a mixture of two solutes is applied to the column as a discrete sample zone occupying a negligible length of the column. The solutes distribute between the carrier gas and the stationary phase according to their respective retention factors. Each solute is then transported along the column by the carrier gas at a rate which is dependent on the retentive effect of the stationary phase.

The retention factor is related to the retention time characteristics by,

$$k = \frac{t_R - t_0}{t_0} \tag{2.20}$$

THEORY OF OPEN TUBULAR COLUMNS

or
$$t_R = t_0(1 + k) \tag{2.21}$$

The mean velocity of any chromatographic zone is obtained by dividing the column length by the retention time of the component, i.e.,
velocity of component = L/t_r

$$= \frac{L}{t_0(1 + k)} \tag{2.22}$$

$$= \frac{1}{(1 + k)} \times \text{mean velocity of carrier gas} \tag{2.23}$$

The **relative** velocity of any two components A and B whose retention factors are k_A and k_B, is consequently $(1 + k_B)/(1 + k_A)$. This is closely related to the separation factor α (1.11) which is simply the ratio of the two retention factors, k_B/k_A.

If k_A and k_B are equal then $\alpha = 1$ and the two components will travel at exactly the same rate and can never separate regardless of the length or plate number of the column. Since retention factors are directly proportional to partition coefficients then (see Chapter 1) the individual migration rates depend on:

1. the vapour pressures of the components at the column temperature;
2. the affinity of the components for the stationary phase.

Therefore compounds with low boiling points (high vapour pressures) will travel faster than compounds with higher boiling points providing condition 2 is similar for all the compounds. The order of elution from the column will then tend to be in order of increasing boiling points. This would be the case for instance where **Raoult's law** applies and the system behaves ideally. If Raoult's law was always applicable, then gas chromatography would only be able to separate compounds having different vapour pressures or boiling points and this would obviously impose a severe limitation on the technique. Fortunately, Raoult's law rarely applies in practice and deviations can easily be invoked by choosing an appropriate stationary phase. Condition 2 is therefore very important since it enables us to choose phases for separations that would otherwise be very difficult.

The effect of the stationary phase on the separation factor, α, can be regarded as the **selectivity** of the system. It is important to realize that the separation factor, and of course the stationary phase selectivity, are completely independent of the column plate number. Any two columns will give the same separation factor for two specified compounds if they contain the same stationary phase and are operated at the same temperature. The two resolutions, however, may differ considerably depending on the theoretical plate numbers of the two columns as it is this property that measures the amount of zone broadening produced by each column. The factors affecting the separation factor will be discussed in Chapter 4 and are referred to as **thermodynamic** effects.

Zone broadening inside the column, otherwise known as **zone dispersion** is largely due to limited rates of diffusion which restrict the rate at which components equilibrate between the two phases. We have seen that the rate at which zone dispersion occurs as the zone travels along the column is measured by the theoretical plate number. Thus high efficiency columns equilibrate very rapidly and so give low zone dispersion and high plate numbers. Consequently the zones are confined to minute regions of the column and emerge as very narrow bands. The chromatograms from such columns are therefore characterized by very narrow and sharp peaks for the components.

To summarize, the overall separation process consists of two distinct mechanisms as follows.

- Zone centres separate at a rate dependent on the selectivity of the stationary phase and the vapour pressures of the components. The separation factor must be greater than unity by an amount which is dependent on the column plate number.
- Zone dispersion occurs at a rate dependent on the **plate number** of the column. Thus, a certain minimum length of column is necessary to produce an effective separation.

These two factors are shown diagrammatically in Figure 2.2 which illustrates the zone concentration profiles for two components at different stages along the column.

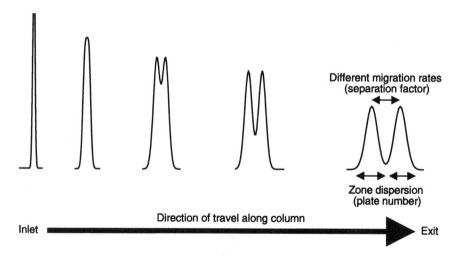

Figure 2.2 Zone concentration profiles at various stages along the column

2.4 PLATE THEORY OF CHROMATOGRAPHY

2.4.1 Discontinuous Model

The plate theory of chromatography was originally derived from a similar concept which was applied to distillation columns by Peters [1]. In their now historical paper on partition chromatography, Martin and Synge developed this idea on the basis that the separation process could be simulated by a number of sequential extraction stages, or theoretical plates. Each of these extraction stages are assumed to contain the two phases in the same relative amount as in the actual column, and within each of these stages there is perfect equilibrium of the solute between the two phases. The analogy here with a real column is that the length of each of these stages, i.e. the **plate height**, H, is a layer in the column of such thickness that the solute concentration in the mobile phase leaving the layer is in equilibrium with its mean concentration in the stationary phase in the layer. Thus, the plate theory is not concerned with the actual nature of the diffusion processes involved, but it assumes that because of the finite mass transfer rates inside the actual column, a situation is produced that can be treated in this way.

Let us assume that a unit mass of a single component is placed in the first plate, and this is allowed to equilibrate completely between the two phases, i.e. after equilibration a fraction a is in the stationary phase and a fraction b is in the mobile phase.

We then transfer the mobile phases in all of the plates to the subsequent ones and again allow equilibrium to establish. Obviously fraction b is now transported to the second plate and again equilibrates between the two phases, as does the fraction a left in the first plate. This process is repeated sequentially until eventually the component vacates the system.

Table 2.1 shows the amount of solute after six transfers.

Table 2.1 Binomial distribution of solute according to the plate theory

Plate volumes of mobile phase	Plate number						
	1	2	3	4	5	6	7
0	$a+b$						
1	a	b					
2	a^2	$2ab$	b^2				
3	a^3	$3a^2b$	$3ab^2$	b^3			
4	a^4	$4a^3b$	$6a^2b^2$	$4ab^3$	b^4		
5	a^5	$5a^4b$	$10a^3b^2$	$10a^2b^3$	$5ab^4$	b^5	
6	a^6	$6a^5b$	$15a^4b^2$	$20a^3b^3$	$15a^2b^4$	$6ab^5$	b^6

We see that the amounts of solute in successive plates can be represented by successive terms in the binomial expansion $(a + b)^n$

Thus, the amount of solute in the $(r + 1)$th plate after the passage of n plate volumes of mobile phase is,

$$Q_{r+1} = \frac{n! a^{n-r} b^r}{r!(n-r)!} \quad (2.24)$$

If the column contains $r + 1$ theoretical plates, and n_{max} plate volumes are required to transport the zone of maximum concentration to the last plate, then $r/n_{max} = 1/(1 + k)$. Also $a + b = 1$, and $a/b = k$ and so $a = r/n_{max}$ and $b = 1 - r/n_{max}$. Both n and r are large in value and approximation formulae can be applied to the general condition described by 2.24, namely,

$$Q_{r+1} \cong \frac{1}{\sqrt{2\pi r}} e^{-\frac{(n_{max} - n)^2}{2r}} \quad (2.25)$$

This is similar in form to the equation to the Gaussian curve of error which can be stated as,

$$y = \frac{1}{\sqrt{2\pi \sigma^2}} e^{-\left(\frac{(\bar{x} - x)^2}{2\sigma^2}\right)} \quad (2.26)$$

where y is the ordinate value for any value of x; $\bar{x}$ is the mean value of x (i.e., the value of x where y is a maximum) and σ is the standard deviation. By comparison we can see that the binomial expansion becomes Gaussian in shape for large values of r and n with a standard deviation value of $\sqrt{r}$ when the zone reaches the end of the column.

2.4.2 Continuous Flow Model

The above treatment is open to criticism because it assumes a multi-stage process similar to the Craig countercurrent system [2]. More accurate treatments based on a continuous flow model have been applied by Gluekauf [3] and by Klinkenberg and Sjenitzer [4].

Let the mass of solute on the rth theoretical plate $= Q_r$. The passage of a volume dV of carrier gas will convey a quantity $dQ(a)$ from the $(r - 1)$th plate to the rth plate, but a quantity $dQ(b)$ will be lost from the rth plate to the $(r + 1)$th plate.

The total change in Q_r is therefore $[dQ(a) - dQ(b)]$.

$dQ(a)$ and $dQ(b)$ are the fractions of the solute in the carrier gas in the $(r - 1)$th and rth plate respectively. These fractions will be $1/(1 + k)$ times the total sample in each plate.

THEORY OF OPEN TUBULAR COLUMNS

Consequently the amount transferred is $1/(1+k)$ times the volume v_g of carrier gas in each plate. Hence

$$dQ(a) = Q_{r-1} \frac{dV}{(1+k)v_g} \qquad (2.27)$$

and

$$dQ(b) = Q_r \frac{dV}{(1+k)v_g} \qquad (2.28)$$

Hence the total change in plate r is given by:

$$dQ_r = dQ(a) - dQ(b) \qquad (2.29)$$

$$= (Q_{r-1} - Q_r) \frac{dV}{(1+k)v_g} \qquad (2.30)$$

The solution to this equation is:

$$Q_r = \frac{\phi^r}{r!} e^{-\phi} \qquad (2.31)$$

where $\Phi = V/(1+k) \cdot v_g$, and V is the total volume of carrier gas required to bring Q_r to the rth plate.

For large values of r we can use Stirling's approximation for factorials, i.e. $x! \cong x^x e^{-x} \sqrt{2\pi x}$ hence:

$$Q_r = \frac{\phi^r e^{-\phi}}{r^r e^{-r} \sqrt{2\pi r}} \qquad (2.32)$$

$$= \frac{\phi^r e^{-(\phi-r)}}{r^r \sqrt{2\pi r}} \qquad (2.33)$$

Q_r is a maximum when $e^{-(\Phi-r)} = 1$, hence $\Phi = r$ at this point, which we will call r_m. Hence we can now write (2.33) as

$$Q_r = \frac{r_m^r e^{-(r_m-r)}}{r^r \sqrt{2\pi r}} \qquad (2.34)$$

or

$$\ln(Q_r) = r - r_m + r \ln(r_m) - r \ln(r) - 0.5 \ln(2\pi r)$$

let $r = r_m + \Delta r$, then

$\ln(Q_r) = \Delta r + r_m \ln(r_m) + \Delta r \ln(r_m) - r_m \ln(r_m + \Delta r) - \Delta r \ln(r_m + \Delta r) - 0.5 \ln(2\pi r)$

$= \Delta r - r_m \ln(1 + \Delta r/r_m) - \Delta r (\ln(1 + \Delta r/r_m) - 0.5 \ln(2\pi r)$

$= \Delta r - (\Delta r - \Delta r^2/2r_m - \Delta r^3/3r_m^2 \cdots) - (\Delta r^2/r_m - \Delta r^3/2r_m^2 \cdots) - 0.5 \ln(2\pi r)$

neglecting the cubic terms, this simplifies to:

$$\ln Q_r \approx -(\Delta r^2)/2r_m - 0.5 \ln(r_m + \Delta r)$$

or

$$Q_r = \frac{1}{\sqrt{2\pi r_m}} e^{-\left(\frac{\Delta r^2}{2r_m}\right)} \quad (2.35)$$

As with the binomial expansion for large n values, the Poisson distribution also approaches that of a Gaussian distribution. Comparison of (2.35) with (2.26) shows that the standard deviation for the Poisson distribution is $\sqrt{r_m}$. This very simple conclusion shows:

1. that the rate of band broadening for any single chromatographic zone in capillary GC is proportional to the square root of the distance travelled by the zone centre. In other words, if N is the number of theoretical plates in the column then when the zone centre reaches the end of the column it occupies $4\sqrt{N}$ plates, assuming that 96% of the solute mass is contained within two standard deviations of the zone centre;
2. that this standard deviation is independent of the amount of solute analysed by chromatography assuming that the partition coefficient remains constant for the solute.

We have seen that $\Phi = r_m$, i.e.

$$\frac{V}{(1 + k)v_g} = r_m \quad (2.36)$$

If $N = $ the total number of theoretical plates in the column, then V is the volume of carrier gas required to bring the zone centre to the end of the column. This is called the **retention volume**, V_R.
Hence:

$$\frac{V_R}{(1 + k)v_g} = N \quad (2.37)$$

or

$$\frac{V_R}{Nv_g} = (1 + k)$$

THEORY OF OPEN TUBULAR COLUMNS

Nv_g is the volume of carrier gas in the column, and the zone travels through this volume whilst a volume V_R passes through the column. Thus the carrier gas travels $(1 + k)$ times faster than the zone centre. This verifies the simple assumption made earlier in (2.23) which was based on the bulk distribution of the zone between the two phases.

2.4.3 Calculation of Plate Number

As the chromatographic zone elutes from the column it is sensed by the detector and produces a Gaussian peak on the chromatogram. Thus, 96% of the peak elutes between $t_R - 2\sigma$ and $t_R + 2\sigma$, where σ is the standard deviation of the peak. This range, measured in units of distance, is **defined** as the **peak width**, as illustrated by Figure 2.3. The standard deviation of the normal error curve is the distance of the points of inflexion from the centre, and so the peak width could in theory be measured by constructing the tangents to the points of inflexion and extrapolating them through the baseline.

The total retention time, t_R is given by (2.21) as $t_R = t_0(1 + k)$. However $t_0 = L/\bar{u}$ and $L = NH$, therefore

$$t_R = \frac{NH(1 + k)}{\bar{u}} \qquad (2.38)$$

also, expressing the peak width w_b in time units,

$$w_b = \frac{\text{length of column occupied by zone}}{\text{zone velocity}} \qquad (2.39)$$

$$= \frac{4H\sqrt{N}(1 + k)}{\bar{u}} \qquad (2.40)$$

hence, dividing (2.38) by (2.40)

$$\frac{t_R}{w_b} = \frac{N}{4\sqrt{N}}$$

i.e.

$$N = 16\left(\frac{t_R}{w_b}\right)^2 \qquad (2.41)$$

Because of the very narrow width of capillary GC peaks it is impracticable to construct peak tangents and to employ (2.41) for the calculation of plate numbers. However an important Gaussian feature is that the ratio of the peak width at any particular fraction of its height to w_b is constant. We can therefore measure the peak width at any defined fraction of its height and employ a

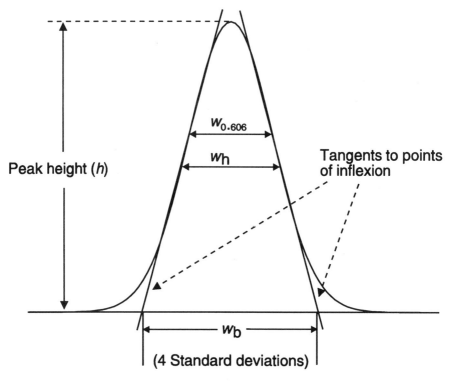

Figure 2.3 Gaussian peak parameters. Key: w_b = peak width; w_h = width at half height; $w_{0.606}$ = width at 0.606 of height

modified version of (2.41) to include the appropriate correction factor. The most popular option was given previously (1.9) and involved the measurement of the peak width at **half height** (w_h). An alternative is to measure the peak width at 0.606 ($w_{0.606}$) of the peak height, which is equal to two standard deviations, or one half of the peak width. The relevant equation in the latter case is,

$$N = 4\left(\frac{t_R}{w_{0.606}}\right)^2 \qquad (2.42)$$

The plate number is dimensionless and so both t_R and $w_{0.606}$ must be measured in the same units. The most convenient method is to measure the equivalent distances as accurately as possible from the chromatogram. The very small peak widths obtained in capillary chromatograms will normally require the use of a magnifier with scale, obtainable from most commercial suppliers. Both (1.9) and (2.42) should, of course, give the same value and so it is irrelevant which of the two methods is used.

THEORY OF OPEN TUBULAR COLUMNS

The theoretical plate number is the usual method of describing the general quality of GC columns without any reference to specific separations. Thus, while it does not indicate the resolving power of the column directly, it does indicate its ability to produce narrow peaks.

The plate number normally increases proportionally with the column length, provided the pressure drop across the column is not excessive. A more specific measurement of column efficiency is the **plate height**, H, obtained by dividing the column length by its theoretical plate number. Hence, a column with a large plate number will give a small plate height. The plate height can be regarded as the mean distance travelled in the column by the zone in order to achieve equilibrium between the two phases. Thus, very rapid equilibration will result in a very small plate height and hence a high column efficiency.

According to the plate theory the standard deviation of the zone at the column exit is given by:

$$\sigma = \sqrt{N} \quad \text{expressed as a number of theoretical plates}$$
$$= H \times \sqrt{N} \quad \text{expressed as a distance}$$
$$= \sqrt{LH}$$

namely,

$$H = \frac{\sigma^2}{L} \tag{2.43}$$

Thus the plate height is the variance of zone dispersion per unit length of column.

2.5 KINETIC, OR RATE THEORY OF CHROMATOGRAPHY

2.5.1 Van Deemter Theory

The diffusion processes occurring in column chromatography were originally studied for packed columns by several workers but particularly by van Deemter and coworkers [5]. The latter authors evolved an equation, popularly known as the **van Deemter** equation which relates the plate height to the column parameters, the carrier gas velocity being the independent variable. Equation (2.44) can be regarded as a statistical relationship involving several summed terms, each of which is a variance contribution to the total zone dispersion.

$$H = 2\lambda d_p + \frac{2\gamma D_g}{\bar{u}} + \left[\frac{8kd_f^2}{\pi^2(1+k)^2 D_s}\right]\bar{u} \tag{2.44}$$

where λ = packing constant; d_p = mean particle size of packing; γ = tortuosity constant due to packing; d_f = mean stationary phase film thickness; D_g = diffusion coefficient of the solute in the carrier gas; D_s = diffusion coefficient of the solute in the stationary phase.

The proposal of this relationship was a major contribution to understanding the nature of the equilibration process in chromatography, and the effects of the variables involved. Expression (2.44) can be stated in its simplified form:

$$H = A + \frac{B}{\bar{u}} + C\bar{u} \qquad (2.45)$$

where each of the three terms is attributed to a particular type of diffusion, or mass transfer process as follows.

The eddy diffusion term A

This is a cause of zone dispersion due to the diverse paths of the solute molecules through the column packing. This term is independent of carrier gas and its velocity, but dependent on the quality and particle size of the packing. There has been some controversy in the use of the term 'eddy diffusion' to describe this contribution, and the alternative **multipath term** is now usually preferred.

Longitudinal diffusion term B

This term describes dispersion of the zone due to molecular diffusion in the carrier gas in an axial direction. Molecular diffusion occurs at a constant rate in both axial directions from the zone centre. The actual amount of dispersion arising from this cause will therefore depend on the mean residence time of the component in the carrier gas. For a column operating under isothermal conditions this residence time is t_0 and is the same for all components because $t_R - t_0$ is the residence time in the stationary phase, and the sum of the two residence times is the retention time of the component. The residence time t_0 therefore depends only on the diffusion coefficient of the component in the carrier gas, the gas velocity, and any obstruction to diffusion by the presence of packing in the column. This effect is illustrated in Figure 2.2 which depicts the zone centred inside the column with the carrier gas turned off (i.e. $\bar{u} = 0$). The zone would then have infinite residence time and consequently would diffuse to fill the column in due course. Conversely, at a high gas velocity the residence time of the zone inside the column would be short and so little longitudinal diffusion could occur. This explains the inverse dependency of longitudinal diffusion on the carrier gas velocity.

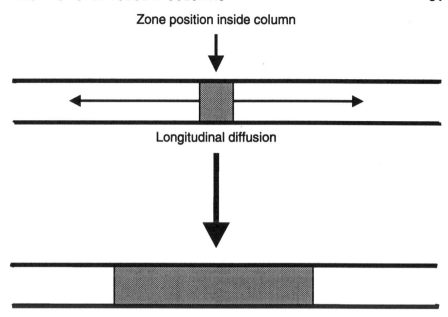

Figure 2.4 Longitudinal diffusion contribution to zone dispersion

Mass transfer term C

This term describes dispersion of the zone due to the distribution of equilibration times for different solute molecules. In the first version of the van Deemter equation, it was anticipated that only whilst molecules were in the stationary phase would they contribute to zone dispersion because of the slow rate of diffusion in liquids compared to gases. Thus, molecules at any point in the liquid stationary phase would be released over a finite time period into the carrier gas phase thereby causing some contribution to dispersion. Solute molecules in the carrier gas were assumed to diffuse sufficiently fast so as to give only a negligible contribution to the overall equilibration process, but the effect of gas phase mass transfer was later included as an additional mass transfer term, C_g, namely:

$$H = A + \frac{B}{\bar{u}} + (C_g + C_s)\bar{u} \qquad (2.46)$$

where

$$C_g = \frac{\psi k^2 d_p^2}{(1 + k)^2 D_g} \qquad (2.47)$$

where ψ = dimensionless constant.

The non-equilibrium processes are termed **resistance to mass transfer** in the gas and stationary phases, and the constants C_g and C_s are the respective **mass transfer coefficients**.

The mass transfer process is illustrated by Figure 2.5.

According to the van Deemter theory the performance of any packed column should be predictable if the various parameters are known. In practice, however, agreement is poor because packed columns have a complex geometry and there is a great deal of uncertainty in quantifying the parameters. For instance, the d_f term has little meaning when the stationary phase is actually distributed as small droplets within a porous structure rather than as a uniform film. Similarly, d_p relates more to uniform spherical particles than to irregularly shaped porous particles. Nevertheless, the practical performance of columns does agree with the trends described by the theory. The effects of the choice of carrier gas, for instance, as described by the D_g value and its velocity $\bar{u}$ are as predicted by the van Deemter equation. Thus **there is an optimum gas velocity which will give a minimum plate height, and this optimum will depend on the choice of carrier gas**. This conclusion has immense practical importance with regard to choosing appropriate operating conditions in general gas chromatography.

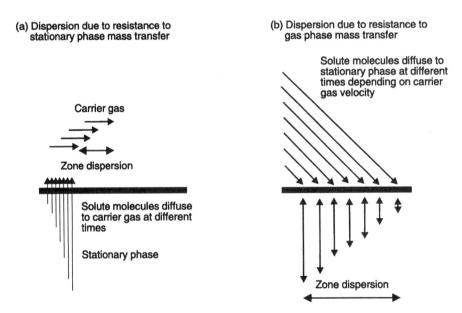

Figure 2.5 Non-equilibrium contributions to zone dispersion

2.5.2 The Golay Theory

The simple and explicit geometry of the open tubular column is a major advantage in relation to any interpretation of its theory because all the parameters, with the exception of diffusion coefficients are either known or can be measured. The equation was first introduced by Golay (see Chapter 1, reference 26) as,

$$H = \frac{2D_g}{\bar{u}} + \left[\frac{(1 + 6k + 11k^2)d_i^2}{96(1 + k)^2 D_g} + \frac{2kd_f^2}{3(1 + k)^2 D_s}\right]\bar{u} \quad (2.48)$$

where d_i = internal column diameter.

Equation (2.48) simplifies to its general form:

$$H = \frac{B}{\bar{u}} + (C_g + C_s)\bar{u} \quad (2.49)$$

Equations (2.48) and (2.49) are similar in form to those for packed columns except for the absence of the A term as this is no longer applicable to an open tube having only a single pathway for the carrier gas.

Golay's proposal to employ coated open tubes for GC in place of packed columns is based on the much higher diffusion rates in gases compared with liquids. Thus, it is not necessary to have the same level of close contact between the two phases as is necessary in liquid chromatography to establish rapid equilibrium. Coating the stationary phase as a thin uniform layer over the inner surface of an open tube provides a situation **at least** as favourable as in packed columns with regard to equilibration speed. This then provides the enormous advantage that resistance to gas flow is reduced by several orders of magnitude, and so much longer columns can be used and much higher plate numbers realized.

Figure 2.6 shows the general nature of equation (2.48) in practice. As with packed columns there is an optimum velocity $\bar{u}_{opt}$ at which a minimum plate height H_{min} will be obtained.

The rapid increase in H as $\bar{u}$ falls below the optimum velocity is due entirely to the effect of the longitudinal diffusion contribution to zone dispersion. Above the optimum velocity there is a gradual loss of efficiency due to the increasing equilibration time between the two phases. The rate at which H increases in this region becomes almost linear with a slope equal to the sum of the mass transfer coefficients $(C_g + C_s)$.

In practice columns should always be operated at or above the optimum gas velocity, and some latitude should be allowed here because different components produce slightly different optima. The practical working range for most columns is normally in the range $\bar{u}_{opt} - 2\bar{u}_{opt}$ (see Section 2.5.7). This working range is generally convenient and appropriate for most applications but it should not be regarded as restrictive. Some columns are used at velocities very

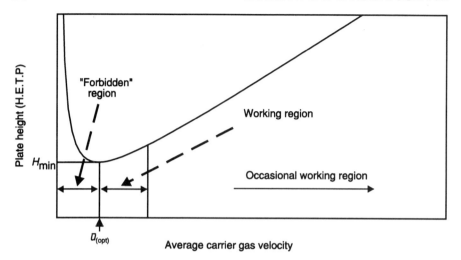

Figure 2.6 Relationship between theoretical plate height and the mean gas velocity according to the Golay equation

much higher than this to accelerate analysis times when the consequential loss of plate number is acceptable. Again, it is the very high permeability of the open tubular column which enables this possibility. **Wide bore columns** (see Chapter 4) are nearly always used at high gas flow rates to achieve fast separations at high sensitivity.

Golay originally considered that resistance to mass transfer in the stationary phase would be the major contribution to non-equilibrium in open tubular columns, as it is with most packed columns. In practice, resistance to gas phase mass transfer is found to be the largest contribution to non-equilibrium in most columns. Slowness of mass transfer in the liquid phase only becomes significant with thick stationary phase films.

The values d_i, d_f, k, and $\bar{u}$ are known accurately for modern columns but there is considerable uncertainty with regard to diffusion coefficient values. According to the kinetic theory of gases, gaseous interdiffusion coefficients are related to molecular parameters and therefore they can vary over a wide range depending on both the solute and the carrier gas. Predictions of D_g can be made via equations such as the Gilliland relationship [6], namely:

$$D_g = \frac{0.0043 T^{3/2}(1/M_1 + 1/M_2)}{p(V_1^{1/3} + V_2^{1/3})^2} \quad (2.50)$$

where T = temperature (K); M = Molecular weights of solute and gas; p = pressure in bar; and V = molar volumes at the boiling point.

The Gilliland equation indicates how D_g depends on the molecular weights, molar volumes of the solute and carrier gas, and the temperature. Of particular

THEORY OF OPEN TUBULAR COLUMNS

importance, also reflected in other relationships is the inverse dependency of D_g on pressure. From the effect of molecular weight we can expect $D_g(H_2) > D_{gg}(He) > D_g(N_2)$ and so the choice of carrier gas will affect all terms containing C_g. Literature data indicates that gas phase diffusion coefficients will lie in the region of 0.1–0.6 cm^2 s^{-1}.

Stationary phase diffusion coefficients are even more uncertain but generally lie in the range 10^{-5}–10^{-7} cm^2 s^{-1}. The use of immobilized and highly polymerized phases creates yet more uncertainty because the stationary phase is no longer a conventional liquid and so it is possible that diffusion into the film does not follow the conventional kinetics of liquid phase diffusion.

A further complication is the effect of the declining pressure along the column. According to (2.50) gaseous diffusion coefficients are inversely proportional to the pressure of the gas and so they will increase as the zone progresses along the column. It follows that the **optimum** velocity also increases but this does not necessarily mean that every part of the column is operating at optimum if the overall **mean** gas velocity is adjusted to the optimum for the column. Consequently if the pressure drop is excessively high there will usually be a gradual change in the plate height from one end of the column to the other, leading to a non-linear relationship between the plate number and the column length. (see Chapter 4). Such extreme conditions are seldom met in practice with capillary columns however, because of their very high permeabilities. Pressure drops are nearly always low and so we can normally expect an almost linear relationship. Giddings [7] has proposed a modified version of the Golay equation which replaces the mean gas velocity $\bar{u}$ by the column exit velocity $\bar{u}_o$, and the gas phase diffusion coefficient becomes the value at the column exit, namely:

$$H = f_1\left[\frac{2D_{g,0}}{u_0} + \frac{(1 + 6k + 11k^2)d_i^2 u_0}{96(1+k)^2 D_{g,0}}\right] + f_2\left[\frac{2kd_f^2}{3(1+k)^2 D_s}\right]u_0 \quad (2.51)$$

where $f_1 = 9/8 \ [(P^4 - 1)(P^2 - 1)]/[(P^3 - 1)^2]$

$f_2 = 3/2 \ [(P^2 - 1)/(P^3 - 1)]$ (the Martin compressibility factor), $P =$ ratio of inlet to outlet pressure, $D_{g,0} =$ Interdiffusion coefficient at the next column exit.

Other workers have proposed similar modifications, particularly Sternberg and Poulson [8], and Ogan and Scott [9].

It is clear that because of the uncertainty in diffusion coefficient values, it is unlikely that the Golay theory could be used for accurate predictions of plate characteristics, but, nevertheless it should be possible to predict trends and relationships of practical significance.

2.5.3 Optimum Values from the Golay Equation

The optimum gas velocity and the equivalent minimum plate height are derived from the Golay equation by differentiation and putting $dH/d\bar{u} = 0$, namely, for minimum plate height H:

$$\frac{dH}{d\bar{u}} = -\frac{B}{\bar{u}^2} + (C_g + C_s) = 0 \tag{2.52}$$

or

$$\bar{u}_{opt} = \sqrt{\left(\frac{B}{C_g + C_s}\right)} \tag{2.53}$$

and

$$H_{min} = 2\sqrt{B(C_g + C_s)} \tag{2.54}$$

A number of important conclusions can be derived from these two simple equations and we shall consider these in relation to practical results to see if the conclusions are supported.

2.5.4 Choice of Carrier Gas

One of the most significant predictions of the Golay theory is that if $C_s \ll C_g$ then (2.53) and (2.54) simplify to:

$$H_{min} = 2d_i\sqrt{2f(k)} \tag{2.55}$$

and

$$\bar{u}_{opt} = \frac{D_g}{d_i}\sqrt{\frac{2}{f(k)}} \tag{2.56}$$

where

$$f(k) = \frac{1 + 6k + 11k^2}{96(1 + k)^2} \tag{2.57}$$

In (2.55) the gas phase diffusion coefficient D_g is no longer present and H_{min} becomes directly proportional to the column diameter d_i and a function of the retention factor k. This indicates that the choice of carrier gas has no effect on the plate number provided the comparison is made at the respective optimum gas velocities for the chosen gases. Also (2.56) shows that the optimum gas velocity is directly proportional to the gaseous diffusion coefficient. **Thus using a lighter carrier gas will increase the speed of analysis.**

Figure 2.7 shows plots derived from the Golay equation after calculating an effective value of the liquid phase diffusion coefficient from experimental values of $\bar{u}_{opt}$ and H_{min} from two columns with different film thicknesses. Assumed values of gaseous diffusion coefficients were applied, namely

THEORY OF OPEN TUBULAR COLUMNS

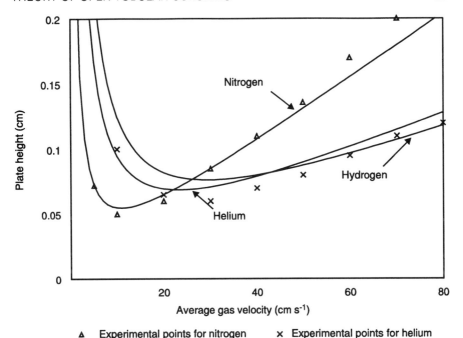

Figure 2.7 Effect of different carrier gases on plate height

D_g(nitrogen) = 0.15 cm^2 s^{-1}, D_g(helium) = 0.4 cm^2 s^{-1} and D_g(hydrogen) = 0.56 cm^2 s^{-1}. The shape of the relationship agrees well with experimental measurements which are included for the named column and for nitrogen and helium.

The column was coated with a 1.2 μm film thickness and gave a loss of about 20% in plate number on using helium in place of nitrogen. This indicates that C_s is having some effect at this level of film thickness which is regarded as a medium thick film.

In general the experimental measurements agree with the theory. Compared with nitrogen, using helium as the carrier gas doubles approximately the speed of analysis and hydrogen gives three to four times the speed. Modern columns are often coated with much thinner films than this and give very little loss of plate number with either helium or hydrogen, showing that C_g is indeed predominant in such cases. Because of the obvious hazards associated with the use of hydrogen, helium is the most popular choice of carrier gas for most capillary columns except those having thick films of greater than about two μm when nitrogen may be the better choice (see Chapter 4).

Figure 2.8 shows chromatograms produced for all three carrier gases, each adjusted to its respective optimum velocity for the particular column. There is

Hydrogen at 47 cm s^{-1}

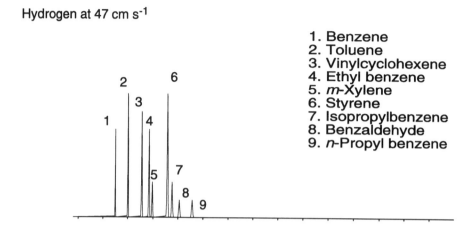

1. Benzene
2. Toluene
3. Vinylcyclohexene
4. Ethyl benzene
5. *m*-Xylene
6. Styrene
7. Isopropylbenzene
8. Benzaldehyde
9. *n*-Propyl benzene

Helium at 30 cm s^{-1}

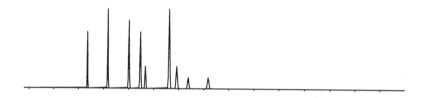

Nitrogen at 15 cm s^{-1}

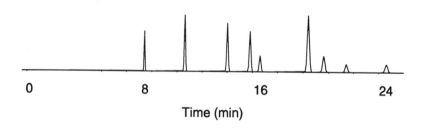

Time (min)

Figure 2.8 Effect of carrier gas choice on separation speed. Column: 50 m × 0.32 mm (1.2 μm) polymethylsiloxane phase (CPSil5CB). (Reproduced by permission of Chrompack International BV)

THEORY OF OPEN TUBULAR COLUMNS

very little loss of resolution as the carrier gas is changed progressively from nitrogen to hydrogen but there is an increasing advantage with regard to speed. A further advantage is the gain in sensitivity with the lighter gases. This is due to the use of the flame ionization detector (FID) which is a mass flow sensitive device (Chapter 3). Detectors of this type give an increase in response with flow rate.

2.5.5 Effects of Film Thickness and Column Diameter

Equations (2.55) and (2.56) give simple predictions of the effects of column diameter d_i and film thickness d_f and we have already seen some of the effects of film thickness in relation to the choice of carrier gas. Equation (2.55) shows that not only is the minimum plate height directly proportional to the

Table 2.2 Experimental theoretical plate characteristics of open tubular columns

Column data: length × diameter (d_f, μm) stationary phase carrier gas, temperature	Plate number N	HETP (mm)	Ratio of plate height to internal diameter
50 m × 0.22 mm (0.2), Cyanopropyl polysiloxane[a] He, 190 °C	218000	0.23	1.05
25 m × 0.32 mm (1.2) Bonded polyethylene glycol N_2, 100 °C	60700	0.41	1.29
50 m × 0.32 mm (0.2) Bonded poly(methylsiloxane)[b] He, 270 °C	152200	0.33	1.03
25 m × 0.25 mm (0.25) Chiral cyclodextrin[c] He, 210 °C	76800	0.33	1.3
25 m × 0.25 mm (0.2) FFAP[d] He, 140 °C	78500	0.32	1.27
10 m × 0.32 mm (0.2) Bonded polyethylene glycol N_2, 60 °C	24500	0.41	1.28
25m × 0.22mm (0.2) Bonded poly(methylsiloxane) H_2, 80 °C	106800	0.23	1.06

[a]Cyanopyopylpolysiloxane = cyanopropyl derivatives of poly(methylsiloxane) (see Chapter 4); a highly polar phase used for the analysis of high molecular weight methyl esters of fatty acids.
[b]Bonded polysiloxane = derivatives of poly(methylsiloxane), cross linked to immobilize them.
[c]Chiral cyclodextrin = a phase incorporating β-cyclodextrin used for the separation of enantiomers.
[d]FFAP = free fatty acid phase, an ester of nitrophthalic acid and polyethylene glycol used for the analysis of free fatty acids.

column diameter (if $C_g \gg C_s$) but its actual value can be calculated from $f(k)$ in (2.57). This approaches a constant value of 0.114 as k increases beyond unity, where H_{min} becomes $0.9d_i$. Allowing for imperfections in manufacture, etc., a useful rule of thumb is that the **minimum plate height is approximately equal to the column diameter**.

Table 2.2 gives some practical values calculated from experimental data. The data does not necessarily relate to optimum operating conditions but it does give an idea of the usefulness of the diameter rule.

With thick film columns it is reasonable to expect that C_s will no longer be insignificant in comparison with C_g and since C_s depends on the *square* of the film thickness we should see a rapidly increasing effect on column performance as the film thickness increases.

Figure 2.9 shows the effect obtained in practice for a very thick film column ($d_f = 5$ μm). Here resistance to mass transfer in the stationary phase is now the predominant factor and so there is an appreciable loss of theoretical plates using helium as a carrier gas compared with nitrogen, as predicted by (2.54). Also we see that the optimum region is very much more critical than with thin

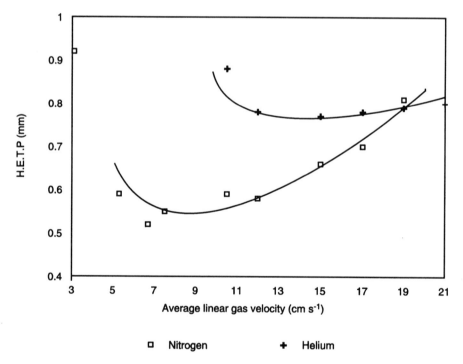

Figure 2.9 Effect of choice of carrier gas on a thick film column. Column: 50 m × 0.53 mm (5 μm) polymethylsiloxane bonded phase. Test compound: dodecane at 140 °C

film columns. It is then preferable to choose nitrogen as the carrier gas in such cases and to operate the column close to its optimum velocity.

The effects of both column diameter and film thickness are illustrated by Figure 2.10. According to (2.48) the plate height is proportional to the square of these parameters which both affect the slope of the linear part of the curve beyond the optimum velocity.

The theory shows clearly that higher plate numbers will be obtained with thin film columns but the overriding factor is the need to ensure that compounds are retained sufficiently by the column to ensure their resolution. For instance with very thin film columns, very volatile compounds may elute close to t_o with little or no retention, and the resulting resolution will be poor irrespective of the plate number. Consequently, it is often necessary to use a thicker film thickness to achieve sufficient retention in spite of the consequential lower plate number.

Figure 2.11 shows the predicted effects of increasing film thickness on the minimum plate heights for nitrogen, helium, and hydrogen as carrier gases.

For small film thicknesses (<1 μm), the plate height depends almost entirely on the column diameter, and as we saw earlier, the choice of carrier gas will have little or no effect on the plate number indicating that the C_g term is predominant. As the film thickness increases, the C_s term becomes increasingly significant and the lines gradually converge showing that the effect of column diameter on plate number is reduced. Ultimately, for thick film columns, i.e. where d_g >2 μm, the diameter of the column becomes almost irrelevant because the plate number then depends almost entirely on liquid phase mass transfer. This is quite important practically because if it is necessary to use such a thick film column to achieve sufficient retention of volatile compounds, then there is no disadvantage with regard to plate number in choosing a wider column than normal. This facilitates some of the practical

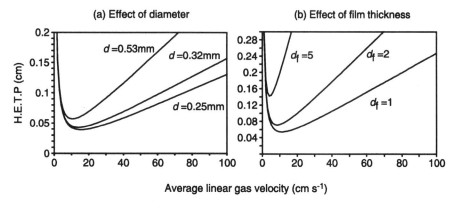

Figure 2.10 Effects of column diameter and film thickness on plate height [1]

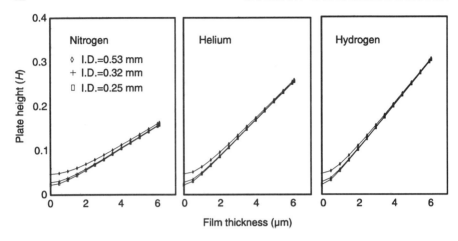

Figure 2.11 Effects of column diameter and film thickness [2]

features of column operation, particularly detection sensitivity and sample introduction. Nevertheless the effect of using a wider column diameter, as predicted by (2.56) still applies and the analysis will be slower.

2.5.6 Effects of Retention Factor on Mass Transfer Coefficients

The mass transfer coefficients C_g and C_s contain functions of the retention factor k, and consequently H_{min} and $\bar{u}_{opt}$ both depend on the choice of component and its retention characteristics. Plate numbers should not be measured on very rapidly eluted compounds as these are not typical of the normal equilibrium situation. For instance unretarded compounds will elute with the carrier gas front, since $k = 0$, where longitudinal diffusion is the major contribution to zone dispersion. Measurements of plate number on these components would give artificially high values. Figure 2.12 shows plots of C_g and C_s versus k for (a) a thin film column ($d_f = 0.1$ μm) and (b) a thick film column ($d_f = 2$ μm).

These plots indicate that measured values of plate number will vary with k whatever the type of column. If C_g is predominant, as for thin film columns, then we can expect to see a gradual decline in the values with increasing k. For thick film columns, where C_s is predominant, the measured column efficiency will show a gradual increase with increasing k. These two examples represent extreme cases and the way that efficiency varies with k in practice will, in general, depend on the balance of the two mass transfer coefficients. Whatever is the case it should be clear that the measurement should only be carried out on components eluting at k values greater than about four.

THEORY OF OPEN TUBULAR COLUMNS

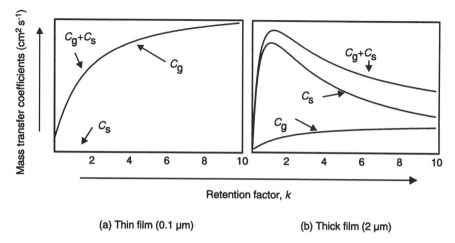

(a) Thin film (0.1 μm) (b) Thick film (2 μm)

Figure 2.12 Effect of retention factor on non-equilibrium coefficients

2.5.7 Practical Operating Range for Gas Velocity

Simplified versions of (2.48), (2.55) and (2.56) are

$$H = \frac{B}{\bar{u}} + C\bar{u} \quad (2.58)$$

$$H_{min} = 2\sqrt{BC} \quad (2.59)$$

$$\bar{u}_{opt} = \sqrt{(B/C)} \quad (2.60)$$

hence

$$\frac{H}{H_{min}} = 0.5 \left[\frac{\bar{u}_{opt}}{\bar{u}} + \frac{\bar{u}}{\bar{u}_{opt}} \right] \quad (2.61)$$

Let $H_r = \frac{H}{H_{min}}$ and $u_r = \frac{\bar{u}}{\bar{u}_{opt}}$, then

$$H_r = \left(\frac{1 + u_r^2}{2u_r} \right) \quad (2.62)$$

This relationship is interesting because it is independent of the column parameters. This implies that provided the general form of the Golay equation is valid then **the relative plate height to the minimum value is dependent only on the carrier gas velocity relative to the optimum velocity!** The significance of this is that the same curve should be obtained for any capillary column regardless of its diameter and film thickness, choice of carrier gas, or

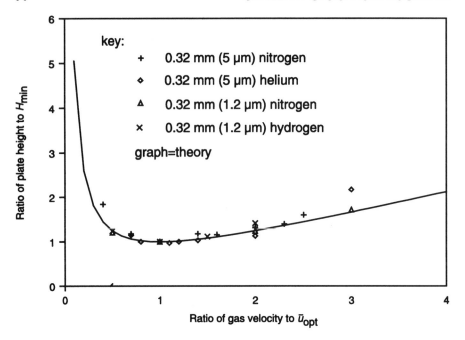

Figure 2.13 Graph of plate heights versus carrier gas velocity expressed as ratios to their minimum and optimum values respectively

the components used for the measurements. In this respect the curve resembles the reduced curves sometimes used in HPLC and first suggested by Giddings.

Figure 2.13 shows the expected theoretical plot of (2.62) as well as experimental points for two different columns and three carrier gases.

The agreement is reasonable and indicates that the basic form of the Golay equation is valid. The major practical conclusion from this is that **the loss of plate height relative to the minimum value** is always the same for similar **relative** velocities. Thus for gas velocities of $2\bar{u}_{opt}$ (i.e. $\bar{u}_r = 2$) the plate height loss is always about 25%. As we shall see in Section 2.6 this will result in only a 10% loss of resolution which is normally tolerable in most applications, justifying the earlier statement that the practical operating range for setting the carrier gas velocity is between $\bar{u}_{opt}$ and $2\bar{u}_{opt}$.

2.6 RESOLUTION

It should be clear from the preceding sections that neither plate number nor separation factor taken individually is suitable for the measurement of resolution. This is illustrated in Figure 2.14 which shows separations carried out

THEORY OF OPEN TUBULAR COLUMNS

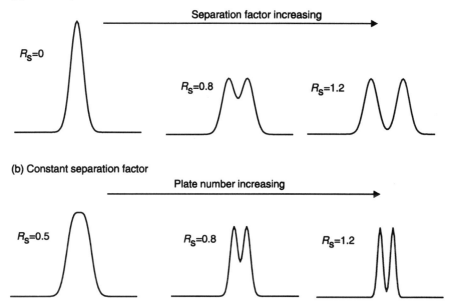

Figure 2.14 Effects of separation factor and plate number on resolution

(a) on different columns having the same plate number but with different separation factors, and (b) on different columns giving the same separation factor but with different plate numbers. The situation depicted in (a) could be obtained from columns containing different stationary phases but otherwise identical from the viewpoint of the column parameters and operating conditions. Situation (b) could result from using columns of increasing length but otherwise containing the same stationary phase and operated under identical conditions.

Resolution must be defined in a way which is relevant to practical analysis, i.e., so that it is related to analytical error when using peak measurement procedures. Naturally any such definition will incorporate the effects of both separation factor and column plate number. Resolution is in fact defined according to the Gaussian nature of the peaks as follows.

$$R_s = \frac{2[t_R(B) - t_R(A)]}{w_b(A) + w_b(B)} \tag{2.63}$$

$t_R(B) - t_R(A)$ is the actual time interval, or distance between the two peak maxima and depends on the separation factor between the two components. The denominator on the other hand, involving only the peak widths, is dependent on the plate number.

Quantification of resolution is essential for two reasons, namely:

1. to provide a minimum criterion for standard methods of test. This is particularly important for referee methods where a minimum resolution should be specified for 'critical pairs' of components
2. to provide a basis for optimization.

Equation (2.63) is based on the Gaussian shape of chromatographic peaks and is a measure of the degree of peak area overlap. Since the peak width w_b is equivalent to 4σ, the resolution according to (2.63) is equivalent to a quarter of the number of average standard deviations between the two peak maxima. **Partially** resolved peak pairs can give rise to various possibilities according to the degree of resolution, namely; a single coherent peak with no visible sign of separation, a peak with a shoulder on its leading or trailing edge, or finally a trough, or minimum between the two peaks. Where a trough is obtained its position and depth will depend on the relative size of the two peaks. Figure 2.15 shows a simple situation where both peaks have the same size and shape and consequently the trough will occur approximately midway between the two peaks. The errors incurred in measuring the areas of the peaks using normal integration methods are shown. For the first peak, part of the tail marked 'b' is lost but the front part of the second peak 'a' is gained. For the second peak, 'a'

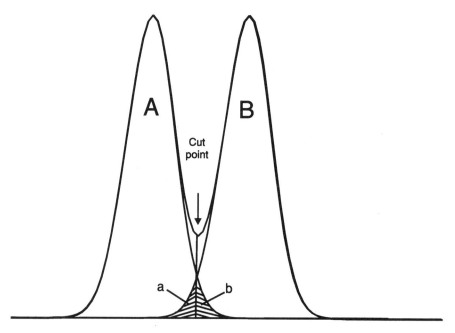

Figure 2.15 Illustration of integration errors from overlapping peaks

THEORY OF OPEN TUBULAR COLUMNS

is lost and 'b' is gained. The total analytical error will, of course, depend on the respective response characteristics of the two components.

If we confine our attention to the errors of area measurement alone then in a simple case of this type, the errors could be obtained from Gaussian tables. In practice, neighbouring components are unlikely to have the same area and the overlap error will depend on the relative amounts of the two components, even if the resolution is maintained constant. Figure 2.16 for instance shows the appearance of two peaks with an increasing amount of peak A relative to B but having a constant resolution of 1.5.

Note how the trough position between the two peaks is pushed away from peak A towards peak B as the relative area of A to B increases. This has the effect of increasing the error on peak B if measured as described above.

Figure 2.17 shows plots of the percentage error on the minor peak area versus resolution for various peak area ratios. For a ratio of 1:1, a resolution of about 1.2 is sufficient to give less than 1% error, but as the ratio increases then greater resolution is required to maintain the error at the same limit.

In order to maintain an error of < 1% then for a peak area ratio of 100:1, $R_s > 1.6$; for 500:1 $R_s > 1.7$ and for 1000:1 $R_s > 1.75$, assuming in all cases that a Gaussian integrity is maintained for both the large and small peaks. However, some peak distortion will often occur for very large peaks due to overloading effects or other disturbances, and some latitude should be allowed for this possibility. For most applications we can limit our attention to ratios of say 1:1 up to 10:1 where a minimum resolution of $R_s = 1.5$ is usually acceptable.

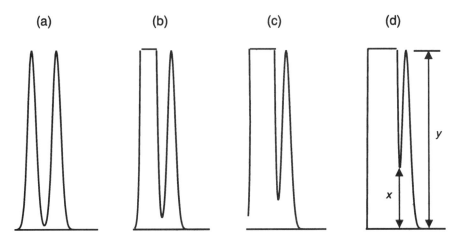

Resolution constant at 1.5: peak area ratios: (a)=1:1, (b)=10:1 (c)=100:1, (d)=1000:1; Trough ratio=x/y

Figure 2.16 Effects of relative peak size on overlap error

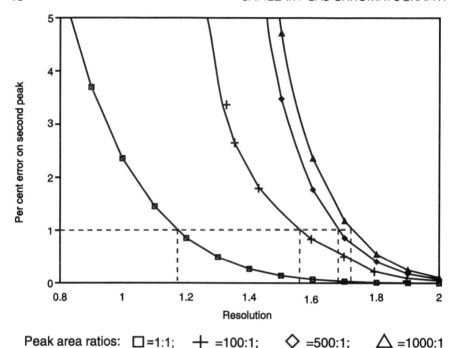

Figure 2.17 Relationship between integration error and resolution for different peak area ratios

Although (2.63) is the accepted definition of resolution, it is not suitable for practical measurements, as discussed previously for column efficiency because of the difficulty of measuring the width of very narrow peaks. The practical method which involves the use of the peak width at half height was given in (1.12). Alternatively the use of the peak width at 0.606 × peak height requires the following expression:

$$R_s = \frac{[t_R(A) + t_R(B)]}{w_{0.606}(A) + w_{0.606}(B)} \qquad (2.64)$$

These formulae can only be used when the two peaks concerned are separated at least to baseline position and from this viewpoint they are satisfactory for use as separation criteria in standard test methods. Another reason for quantifying resolution is for optimization purposes (see Chapters 4 and 8), where components are not initially separated by the required amount. If components co-elute, or give shoulders or troughs, there would be no indication of the action necessary to improve the separation, even if the resolution could be estimated by the use of these formulae.

THEORY OF OPEN TUBULAR COLUMNS

For the purposes of optimization, (2.64) can be transformed into a more useful form as follows:

Since $w_b(A) = 4t_R(A)/\sqrt{N}$ and $w_b(B) = 4t_R(B)/\sqrt{N}$

$$R_s = \frac{[t_R(B) - t_R(A)]\sqrt{N}}{2[t_R(A) + t_R(B)]} \quad (2.65)$$

but $t_R(A) = t_0 + t'_R(A)$ and $t_R(B) = t_0 + t'_R(B)$ where $t'_R(A)$ and $t'_R(B)$ are the *adjusted* retention times.
Therefore

$$R_s = \frac{[t'_R(B) - t'_R(A)]\sqrt{N}}{2[2t_0 + t'_R(A) + t'_R(B)]} \quad (2.66)$$

and dividing throughout by t_0

$$R_s = \frac{[k(B) - k(A)]\sqrt{N}}{2[2 + k(A) + k(B)]} \quad (2.67)$$

For partially resolved peaks eluting very close together, as in capillary GC

$$k(A) \approx k(B)$$

let

$$k_{av} = (k(A) + k(B))/2$$

where k_{av} = average capacity factor for compounds A and B; therefore

$$R_s \cong \frac{[k(B) - k(A)]\sqrt{N}}{4(1 + k_{av})} \quad (2.68)$$

dividing throughout by $k(A)$

$$R_s = \frac{\sqrt{N}}{4}(\alpha - 1)\left[\frac{k(A)}{1 + k_{av}}\right] \quad (2.69)$$

Equation (2.69) is one form of a relationship known as the **resolution equation** and its value in practical optimization cannot be over emphasized. It is often quoted in slightly different forms depending on the approximations employed, but for closely eluting components the differences become negligible. The three factors, α, k, and N are all easily measured, even for partly resolved peaks, and so resolution can usually be calculated. More significantly however, we can now see the effects of changing these variables on the ultimate resolution. These effects are very different from each other, as is discussed in detail in Chapter 4. For instance, increasing the plate number N by using a longer column is the weakest effect, with the resolution only

Table 2.3 Use of trough ratios as measurements of resolution

Area ratio	Trough ratio	Resolution	Fractional Position of trough between peaks	Per cent error
1:1	0.5	0.83	0.5 (half way)	5.22
100:1	0.5	1.26	0.72	8.17
1000:1	0.5	1.41	0.76	8.48
10 000:1	0.5	1.53	0.78	9.2

increasing in proportion to the square root of the column length. For this reason column lengths should only be increased as a last resort, and then by at least a factor of two. If this option is chosen it will have a very deleterious effect on analysis time and detection sensitivity. The factor $k/(1 + k_{av})$ is only effective for low retention factors, i.e., when components elute close to the holdup time. For higher values of k this factor approaches unity and therefore becomes almost ineffective. The most critical factor in this expression is the separation factor α which depends mainly on the choice of the stationary phase.

2.6.1 Trough Ratios as a Measure of Resolution

Another way of quantifying the resolution of partially resolved components is to use the **trough ratio**, calculated as shown in Figure 2.16. The value is zero for baseline resolution but there are many practical situations where baseline resolution is unattainable. For instance, chromatograms from trace analysis often show the presence of components in the tail of major components. [Figure 2.16(d)]. Measurement of the areas of such minor components may involve various ways of situating the baseline position but the method should also include an appropriate separation criterion, as mentioned earlier, and the use of the trough ratio can be useful for this purpose.

Computer analysis was applied to simulated situations to calculate the error on the minor peak for various peak area ratios but with a constant trough ratio. The results are given in Table 2.3. Although the necessary resolution increases as the area ratio increases, the error is approximately the same for peak area ratios of 100:1 up to 1000:1.

2.7 GENERAL SUMMARY

The general theory of capillary GC provides a number of very useful guidelines for the practical operation of the technique. The vast majority of modern columns are coated with thin to medium films of stationary phase in the range 0.1–1 μm. Thicker films should be selected for special cases, principally to provide sufficient retention for volatile compounds but occasionally to minimize adsorptive effects from the column wall. The major conclusions are listed as follows.

1. Thin film columns give better plate numbers than thick film columns and operate at faster optimum gas velocities.
2. Most columns should be operated either with helium or hydrogen carrier gas. Hydrogen gives the fastest analysis but helium is often preferred for safety reasons. Helium is about twice as fast as nitrogen. Neither helium or hydrogen will give a significant loss of plate number if the film thickness is less than about 1 μm.
3. Thin film columns should give a plate height approximately equal to the column internal diameter.
4. Thick film columns (where $d_f > 2$ μm) will give some loss of theoretical plates with lighter carrier gases. If plate number is to be maximized then it is better to use nitrogen at close to its optimum velocity. This becomes very important in the case of very thick films.
5. The column diameter affects the speed of the column since u_{opt} is proportional to $1/d$. Most contemporary columns are 0.2–0.3 mm in diameter but smaller diameters may be used if a higher speed is required. Speed is seldom a problem in capillary GC however, as it is already a very fast technique and it is more common in fact to choose wider bore columns in view of their greater capacity and other advantages which are discussed in Chapter 4.

REFERENCES

1. Peters, K. (1922) *Industr. Eng. Chem.*, **14**, 476.
2. Craig, L. C. (1944) *J. Biol. Chem.*, **155**, 519.
3. Gluekauf, E. (1955) *Trans. Faraday Soc.*, **51**, 34.
4. Klinkenberg, A. and Sjenitzer, F. (1956) *Chem. Eng. Sci.*, **5**, 258.
5. van Deemter, J. J., Zuiderweg, F. J. and Klinkenberg, A. (1956) *Chem. Eng. Sci.*, **5**, 271.
6. Gilliland, E. R. (1934) *Industr. Engng. Chem.*, **26**, 681.
7. Giddings, J. C. (1964) *Anal. Chem.*, **36**, 741.
8. Sternberg, J. C. and Poulson, R. E. (1964) *Anal. Chem.*, **36**, 58.
9. Ogan, K. and Scott. R. P. W. (1984) *J. High Res. Chromatog.*, **7**, 382.

3
Capillary Instrumentation

3.1 INTRODUCTION

The modern capillary gas chromatography can take a number of different forms depending on its commercial origin and intended field of use. Whatever is the nature of the instrument, it represents the culmination of several decades of research and development. The enormous application range of the technique means that adaptability and versatility are important requirements.

As discussed in the introduction, the first generation of chromatographs was designed for packed columns and had a basic construction involving simple flow control systems and a limited range of detectors. The entire instrument was usually designed as an inflexible single channel unit in which detectors, injectors, etc., could not be easily changed or upgraded.

The introduction of open tubular columns placed stringent demands on instrument design and the arrival of new detectors and sample introduction techniques led to the appearance of separate, interchangeable and replaceable units. This inevitably led to hybrid systems, which, although satisfactory in practice, created a situation that was not in the best interests of manufacturers.

Instrument design has in most cases returned to an integrated format but now offers a choice of options including various injectors, detectors, valving, data systems, etc. The data system is sometimes a separate unit but it may be integrated with the chromatographic system. Whatever the make of the instrument one important requirement is that it should be possible to install columns from any source, although this may involve the use of different fittings. Most instruments can also accommodate at least two columns, used either independently or coupled in some way.

Perhaps the most significant development in recent years has been the introduction of computing systems and software for data collection and processing. Data systems are often regarded as essential to modern chromatography, particularly in busy laboratories where a large number of analyses are performed on a routine basis.

There is no doubt that the modern instrument presents an impressive image to the novice, in contrast to that presented by the column itself, which many might regard as simply an appendage! A common misconception is that expensive instrumentation coupled with a sophisticated data system will make up for any deficiencies in the quality of the column. In fact one of the most important messages to the potential user of GC is to understand that the purpose of the instrument is to service the column in the most effective way, i.e., to provide the column with its requirements. **The ultimate quality of the separation depends on the quality of the column!**

3.2 CAPILLARY COLUMN REQUIREMENTS

The basic instrumental requirements are depicted as five major facilities, i.e.,

- control of column temperature;
- supply and control of carrier gas;
- means of sample introduction;
- detection of eluted components;
- data handling and analysis.

At first sight these requirements appear to be fairly undemanding, but the minute size and scale of the capillary column has been the cause of many problems in design and technique, some of which are still of concern. A critical feature is the requirement that the instrument itself should be designed and operated so as not to cause additional broadening of the chromatographic peaks. Such **extra-column** effects broaden the final peak width according to the following variance law:

$$\sigma_{tot}^2 = \sigma_{col}^2 + \Sigma \sigma_{ex}^2 \qquad (3.1)$$

where σ_{tot}^2 = variance of final peak; σ_{col}^2 = variance of column zone dispersion processes; $\Sigma \sigma_{ex}^2$ = sum of extra-column variances. These extra-column variance arise from three principal areas, namely,

$$\Sigma \sigma_{ex}^2 = \sigma_{det}^2 + \sigma_{inj}^2 + \sigma_{conn}^2 \qquad (3.2)$$

where σ_{inj}^2 = variance of sample introduction technique; σ_{det}^2 = variance of detection technique; σ_{conn}^2 = variance due to connections.

The variances in (3.1) can be quoted in any consistent units, but it is most useful to consider the effects in volumetric units, namely,

$$V_p^2 = V_{col}^2 + V_{inj}^2 + V_{det}^2 + V_{conn}^2 \qquad (3.3)$$

V_p is the peak volume, i.e., the volume of carrier gas which elutes from the column during the time period equivalent to w_b, the 'width' of the peak. V_{col} is the 'ideal' peak volume, i.e., the peak volume that would be obtained in a perfect instrument and arising from column zone dispersion processes alone. V_{inj} is the volume of carrier gas and sample vapour involved in transporting the sample to the column; V_{det} is the effective volume of the detector, and V_{conn} is the effective volume of the column end connections, or other junctions in the sample flow path. Clearly the peak volumes associated with capillary columns are very small quantities, because

$$V_p = w_b F_{av} \quad (3.4)$$

Where F_{av} is the mean carrier gas volumetric flow rate calculated from the mean carrier gas velocity and the cross-sectional area of the column, i.e.

$$F_{av} = \bar{u}\pi d_i^2/4 \quad (3.5)$$

and

$$w_b = 4t_R/\sqrt{N} \quad (3.6)$$

therefore

$$V_p = \pi d_i^2 \bar{u} t_R/\sqrt{N} \quad (3.7)$$

A 0.25 mm ID column operating with helium carrier gas at 25 cm s^{-1}, and having a theoretical plate number of 100 000 theoretical plates gives a peak volume of 0.093 ml for a component, $t_R = 10$ minutes. Clearly the sum of the extra-column variances, $\Sigma\sigma_{ex}^2$, must **together** be rather less than this for the column to be operated efficiently.

Equations (3.2) and (3.3) illustrate several important features that **must** be addressed if successful capillary operation is to be achieved. One problem that is not immediately apparent from the variance law is the **type** of extra-column volume which can lead to additional zone broadening. In simple terms there are two effects, namely, that arising from the simple geometrical volume of instrument components, and 'stagnancy' effects arising from poor instrument design, the use of an inappropriate system which is not compatible with the column selected, or from poor installation of the column. Extra-column volumes can arise from any or all of these effects in combination. For example, Figure 3.1 shows some common effects as they may apply in practice. In (a), on the injector side of the column, the space between the split point and the top of the column is only purged slowly and so acts as a semi-stagnant region where the sample concentration will slowly decline over a period of time. This will result in tailing peaks for the solvent and sample components on the chromatogram. In (b) where the column is correctly installed, with the appropriate injector design, any semi-stagnant regions are rapidly purged and so sharp peaks are obtained.

The detector contribution to peak broadening, represented by V_{det} in (a) is mainly due to its geometrical volume, although there may also be stagnant regions if the geometry is poorly designed. Generally the physical volume of

CAPILLARY INSTRUMENTATION

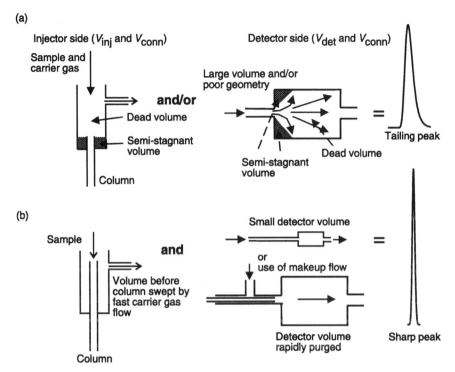

Figure 3.1 Illustration of extra-column volume effects arising from the injector and detector. (a) Poor installation and/or design; (b) correct installation and/or design

the detector must be very small in comparison with capillary peak volumes. This is not always possible, and some detectors require a subsidiary flow of gas, known as a **makeup** flow which is introduced at a point between the column and the detector. This additional gas flow purges the detector rapidly and reduces its effective volume.

The extra-column volume contribution from connections, V_{conn}, also consists of geometrical and stagnant volumes. The effects of these unwanted volumes must be minimized by the use of appropriately designed fittings, by ensuring that the fittings are installed correctly, and by avoiding the use of worn or damaged ferrules.

3.3 CHARACTER OF MODERN INSTRUMENTATION

As mentioned earlier, the modern instrument can take many forms depending on its intended use. Although this book is mostly concerned with the **general**

CAPILLARY GAS CHROMATOGRAPHY

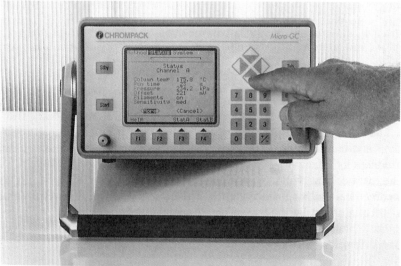

Figure 3.2 (a) Photograph of Shimatzu GC-17A gas chromatograph. (Reproduced by permission of Dyson Instruments Ltd.) (b) Photograph of micro-GC. (Reproduced by permission of Chrompack International BV)

CAPILLARY INSTRUMENTATION

laboratory instrument, capillary GC systems are also available for industrial plant monitoring and control, for environmental monitoring where a degree of portability may be required, and for specific analyses. For the latter, the instrument may be combined with ancillary equipment such as automatic sample introduction, a data system and may also be completely automatic in operation. An instrument of this type is usually called an **analyser** providing a 'turnkey' solution with the operator only being required to perform fairly basic operations.

Figure 3.2a shows a state-of-the-art laboratory chromatography for general use equipped with autosampler. In contrast Figure 3.2b shows a completely portable analyser intended for the very rapid analysis of volatile samples such as natural gas. The latter employs short, small diameter packed and open tubular columns to complete analyses within minutes for field and laboratory monitoring.

3.4 THE GAS SUPPLY SYSTEM

It is essential for the carrier gas supply to the column to consist of a pure oxygen-free gas at the appropriate flow rate. The control system must be capable of maintaining the required flow or pressure conditions for the chromatographic run and independently of any auxiliary gases requiring separate adjustment. These auxiliary gases are usually necessary in capillary GC for the operation of injectors and detectors.

3.4.1 Quality and Source of Gas Supplies

Several gas supplies are usually needed to operate a capillary GC system, namely the carrier gas, hydrogen fuel gas for flame detectors, clean air to sustain the flame combustion, and makeup gas if required. All of these gases can be supplied from gas cylinders via the normal cylinder head pressure control valves to reduce the pressure to 5–6 bar. The pressure supplied to the instrument can then be adjusted to whatever values are recommended by the instrument manufacturer. Commercial generators offer a convenient and attractive alternative for the hydrogen and air supplies. They are generally safer and certainly more convenient than gas cylinders, and are capable of supplying several instruments simultaneously.

Whatever the source the various gas supplies must be 'pure'. In the case of the carrier gas it is essential for oxygen, moisture and organic impurities to be absent. Oxygen, in particular, must be absent from the carrier gas to avoid any oxidation of the minute quantity of stationary phase in the column. Any oxidation of the phase will cause changes of polarity followed by a loss of phase from the column with a consequent loss of column performance. High purity gases should always be used of course, but it is essential to include a moisture filter followed by an oxygen filter in the supply carrier gas line. These should consist of high quality commercial filters to reduce the residual oxygen content to less than 1 ppm. Often such filters will have a colour indicator and this should be examined frequently.

Flame detectors are ubiquitous in capillary GC, and require auxiliary supplies of hydrogen and air. All auxiliary gases should be cleansed by charcoal filters to remove any traces of organic impurities. The use of air compressors is not recommended for capillary operation unless they form part of a custom gas supply system for GC, properly stabilized and filtered.

Other practical recommendations are as follows.

1. Replace gas cylinders whilst still partially full, i.e. with 10–15 bar residual pressure.
2. Ensure that the cylinder head fitting is dry and free from any grease, and is in good condition.
3. Use either stainless steel or annealed copper tubing for plumbing. Ensure that all tubing has been washed with solvent to remove grease, etc., and then thoroughly dried. Polymer tubing should not be used as most gases, particularly helium and hydrogen can diffuse through the tube wall. Helium can diffuse out of the system and oxygen can diffuse in!
4. Use good quality compression fittings for any connections and always check for leaks.
5. Ensure that the system can be isolated to prevent any ingress of air when cylinders or filters need to be changed. This will generally require the use of a separate shut-off valve near the supply cylinder. Some commercial filters are designed to automatically prevent any ingress of air when they are changed.

3.4.2 Gas Control Devices

Until recently capillary columns were always operated at **constant pressure**, i.e. so that the pressure drop across the capillary column was held constant. Under these conditions, however, it is impossible to maintain a constant flow rate, and hence constant gas velocity through the column if it is subjected to flow changes for any reason. Columns are frequently operated under **programmed temperature** conditions, in which the column temperature is increased according to a defined profile during the run. The reasons and benefits of this mode of operation are discussed in Chapter 8 but there are inherent problems with this because of the consequential change in the carrier gas viscosity. Gas viscosities **increase** with increasing temperature at about 0.2–0.3 per cent per degree. For a column operating at a constant pressure drop this causes a gradual decrease in the flow rate during the programme which can amount to as much as fifty per cent over a wide temperature programme, resulting in a longer analysis time, reduced detector response, reduced column lifetime, and a loss of theoretical plates if the linear gas velocity falls below the optimum. The recommendation (see Chapter 2) of using a carrier velocity of up to twice the optimum will tend to compensate for these effects however, and so there should not be any serious consequences of constant pressure operation.

There are essentially five different types of gas flow control device used in GC instrumentation, namely, on–off valves, forward and back-pressure regulators, needle valves, and mass flow control regulators. Simple on–off, and needle valves are well known and although used in conjunction with other devices, they do not themselves provide any actual control of flow or pressure. Most commercial systems employ these devices either singly or in combination to accomplish the necessary control of the instrument. Diagrammatical forms of (a) a pressure regulator and (b) a mass flow control device are shown in Figure 3.3.

Pressure regulator

Gas is supplied to the inlet port at a pressure p_1 which should be considerably higher than p_2, the pressure supplied to the column. The manufacturer's recommendations should always be followed here but typically the supply pressure should be in the region of 3–6 bar. The gas flows through the poppet valve attached to the lower side of the stainless steel diaphragm which takes up an equilibrium position depending on the balance of the gas pressure and the adjustable spring pressure. Any tendency for the pressure p_2 to change will change this equilibrium position thereby changing the position of the poppet valve. If, for instance, the pressure p_2 falls, the consequent diaphragm movement allows the valve to **open** slightly allowing more gas to enter so restoring the pressure to its original value.

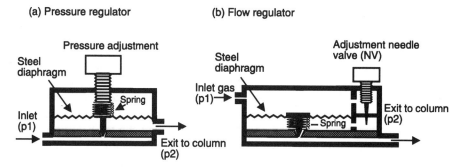

Figure 3.3 Schematic diagrams showing operational principles of (a) a pressure regulator and (b) a mass flow controller, as used in capillary GC

The use of a constant pressure regulator can be a very effective method of control and is independent of the supplied pressure p_1 providing that it is considerably higher than p_2. Since the column is thermostatically controlled then a constant flow rate will be achieved through the column provided it is operated isothermally. However, it is particularly important to avoid small leaks with pressure control as these may not show up from retention characteristics. As gas leaks can have a serious effect on chromatographic performance a leakage check should always be carried out when a new column is installed.

Flow regulator

In Figure 3.3b the flow rate is set by adjusting the needle valve, NV. Here a constant pressure difference is maintained across the valve by the built-in fixed pressure controller which functions as described earlier. This ensures that a constant mass flow rate emerges from the device. For instance, any tendency for the flow rate to increase (via leaks, opening of valves, etc.) will cause the pressure p_2 to decrease. This in turn would cause the pressure on the underside of the stainless steel diaphragm to decrease and hence to lower its position, closing the poppet valve slightly. Thus the pressure will build up again to maintain the same pressure difference across the needle valve, consequently restoring the original volumetric flow rate. The nett effect is that any tendency for the flow rate to change is compensated by an appropriate change in pressure p_2. As mentioned earlier, most capillary columns are not operated with simple flow regulators because of the need to adjust subsidiary gas rates from the same supply independently. Nevertheless most commercial systems do use them in conjunction with pressure controllers and other valving to achieve the requisite flow conditions in all parts of the instrument.

Figure 3.4 shows a typical general purpose flow system for split injection involving the use of various types of regulator. Most commercial systems are variations of this design, particularly when equipped with so-called split–splitless injectors (see Chapter 6).

Important terminology with which the user should become familiar is included in Figure 3.4. The **split flow** refers specifically to the volumetric flow rate through the filter while the **split ratio** is the ratio of this split flow rate to the mean column flow rate. Most sample injectors are designed to include a **septum flush** operating at a few millilitres per minute to avoid column contamination from septum volatiles. The system supplies the **flow controller** (FC) on the inlet side of the injector with a constant pressure P_1 by means of the **pre-pressure regulator** (PPR). The pressure P_2 at the split point is held constant by the **back pressure regulator** (BPC) irrespective of the split flow rate which is adjusted by the needle valve (NV1) and flow controller (FC) settings. Thus, the back pressure regulator ensures a constant column head pressure while the flow controller ensures a constant split ratio. Needle valve

CAPILLARY INSTRUMENTATION

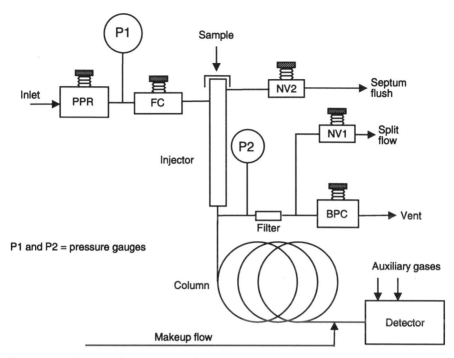

Figure 3.4 Typical flow scheme for capillary systems using split injection

NV2 adjusts the septum flush, and the detector makeup flow may be controlled by needle valve, pressure regulator, flow regulator, or even by a fixed constriction, depending on the type of detector and identity of the makeup gas.

Another requirement of most instruments is a knowledge of the pressure at various points in the flow pathway, particularly the auxiliary gas pressures and the column inlet pressure. Bourdon or similar pressure gauges are commonly employed for measuring pressure in GC, but these are now being replaced by digital gauges involving electronic pressure sensing. In the latter event the pressure readout may be handled by the data system and output to a PC monitor with other data, or possibly to a data screen on the chromatography itself. Much of the flow regime in capillary instrumentation is closely associated with the requirements of sample injection, and will be discussed in more detail in Chapter 6.

As instrument design has evolved so we have seen an increasing use of keypad setting of chromatographic conditions, either from the chromatography itself or from a separate data system. This has been normal for some time as a way of setting temperature and electronic conditions, but gas flow settings are still usually adjusted manually. This can be one of the most difficult aspects of instrument operation for the inexperienced user as it does involve an understanding of the flow system and the specific function of the various controllers and pressure gauges. This situation is beginning to change and several commercial instruments are now equipped with automatic flow regulators which allow column and injector flow conditions to be set digitally via the key pad. This introduces a new dimension in reliability, simplicity of use, and improved column performance as it also enables automatic pressure of flow programming to be performed.

Figure 3.5 is a diagram of a Shimatzu system involving automatic flow control which enables all the useful methods of sample introduction to be employed.

This arrangement is designed to operate in a variety of sample injection modes. The automatic flow control module functions either as a mass flow control or pressure control device and can be programmed to compensate for viscosity changes during programmed temperature operation. The upstream flow control section contains a flow sensor which controls the total flow rate by a feedback circuit. The pressure sensor similarly controls the column pressure. The three-way solenoid valve allows the split flow rate to be turned on and off for splitless injection; otherwise the system is similar to that in Figure 3.4.

Some of the effects of using an automatic system of flow control are shown in Figure 3.6.

The three chromatograms in Figure 3.6 are sections taken from actual chromatograms obtained using Hewlett Packard equipment. Simple setting of

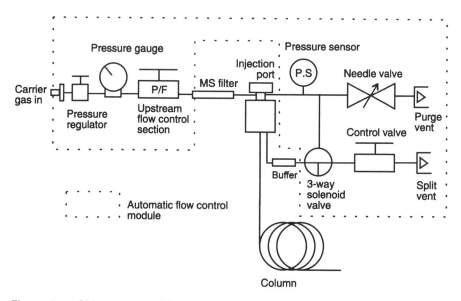

Figure 3.5 Shimatzu GC-17A autoflow system utilizing automatic flow control and adjustment. (Reproduced by permission of Dyson Instruments Ltd)

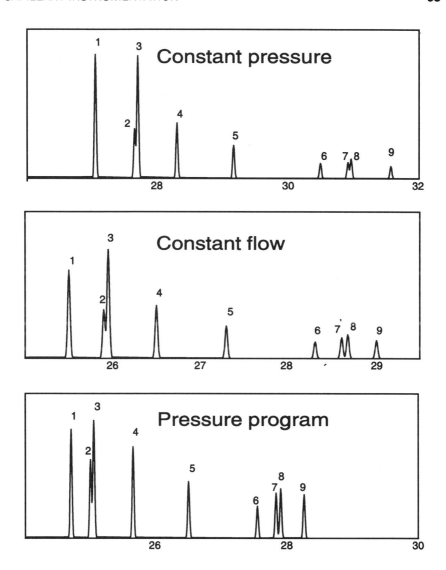

Figure 3.6 Use of automatic flow programming. Peak identities: 1 = di-n-octyl phthalate;   2 = benzo(b)fluoranthene;   3 = 7,12dimethylbenzo(n)anthracene;   4 = benzo(a)pyrene;   5 = 3-methylcholanthrene;   6 = dibenzo(a,j)anthracene;   7 = inden(1,2.3-cd)pyrene;   8 = dibenzo(a,h)anthracene;   9 = benzo($g,h,i,$) perylene. Chromatogram 1: constant pressure at 14 psi; chromatogram 2: constant flow mode 14 psi at 35 °C; chromatogram 3: pressure programme 14 psi (0 min) to 50 psi at 2 psi min^{-1}. MSD; inlet: split/splitless in splitless mode; oven program: 35 °C (1 min) to 310 °C at 10 °C min^{-1}. (Reproduced by permission of Hewlett-Packard Ltd)

the flow conditions is ensured via the control system which allows various options as shown. In chromatogram 1, the effect of the decrease in flow rate due to the viscosity increase when the pressure is constant is seen. In chromatograms 2 and 3 the pressure is programmed to produce respectively, a **constant** flow and an **increase** in the flow rate during the runs. In both of the latter cases the run time was shortened and the heights of later peaks increased. This increase in the chromatographic response is a consequence of using a mass flow sensitive detector, (see Section 3.8.1) which in this case is the Hewlett Packard mass selective detector.

The use of **pressure programming** is not as yet a widely used technique in capillary GC and some experts would question the wisdom of introducing yet another variable into the system, arguing that temperature programming would produce a similar effect. Programmed temperature operation does degrade the separation factor however, whereas pressure programming has no effect on this although there will be some loss of column plate number as the gas velocity moves away from the optimum. The loss of resolution resulting from this effect, however, will usually be less than that resulting from a reduced separation factor. Although this would appear to give pressure programming an advantage over temperature programming, the amount of chromatographic contraction available would be far less because of the logarithmic relationship between temperature and retention factor. Nevertheless, pressure programming is a useful additional facility which is available in some instruments. However, the technique should be used with care and with a proper understanding of the effects.

3.4.3 Measurement of Flow Rates

The volumetric flow rate through the capillary column can be computed from the mean linear gas velocity as described in Section 3.1, and this can often be done by the data system as part of the calculation for determining the split ratio. The split ratio determines the amount of sample entering the column, which, in turn, affects the amount of sample dilution necessary and/or the volume to be analysed by chromatography. These factors must be adjusted to ensure that the amounts of each component do not cause overloading and peak distortion effects.

Other flow rates that need to be measured and set to the appropriate values are detector auxiliary gases, makeup gas, septum purge and split flow rate. Depending on the make and nature of the instrument these may have to be measured using a separate flowmeter. Soap bubble flow meters can be employed but digital meters which give an instantaneous readout are more convenient. Some of the auxiliary gases such as flame detector gases, septum purge and makeup flows may be preset by the manufacturer. Where applicable, the operator would be well advised to set flow rates strictly according to manufacturer's recommendations.

3.5 COLUMN COUPLING AND CONNECTIONS

Satisfactory coupling of the capillary column to the instrument was initially a very difficult aspect of capillary GC, particularly when most columns were of glass construction. Columns are now installed by means of compression fittings of various types. It is important that the fittings are easy to connect and disconnect repeatedly, and of course they must not introduce any significant extra volume to the system. Figure 3.7a shows the principle of the compression fitting, although different manufacturers frequently employ specific fittings for their own equipment.

A critical component of the compression fitting is the **reducing ferrule**. Metal columns will normally use stainless steel for this component, swaged onto the outside of the column when the fitting is first tightened to provide a leakproof seal. This ferrule usually consists of two parts; a **front** and a **backing ferrule**. As most capillary columns are now constructed from fused silica, **soft** ferrules must be used, and these are fabricated from a variety of polymers, graphite or combinations of the two. Table 3.1 is a summary of the properties of the materials used for reducing ferrules.

It is vitally important to observe some basic rules in relation to the use of compression fittings in capillary GC, summarized as follows.

1. Never used worn or distorted ferrules or fittings.

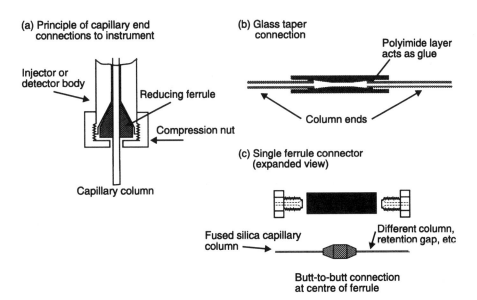

Figure 3.7 Connections used in capillary GC

Table 3.1 Ferrule materials used with capillary columns

Material	Stainless steel (high grade SS)	PTFE Teflon (fluoropolymer)	Vespel (polyimide) graphite–vespel (polyimide + carbon)	Kalrez (a modified PTFE) graphite–Kalrez (fluoropolymer + carbon)	Graphite (pure carbon)
Use with fused silica columns?	No	Yes	Yes	Yes	Yes
Use with metal columns?	Yes	Yes	Yes	Yes	Yes
Reusable?	Yes	No	Yes	Yes	Yes/No
Maximum temperature (°C)	Any	200–500	350 (vespel) 400 (graphite–vesp)	300–320	450
Bleed?	None	Increases rapidly above 200 °C	Only near maximum temperature	Only near maximum temperature	None

Teflon, Vespel and Kalrez are trademarks of E. I. Du Pont

2. Ensure that the correct ferrule material and size are chosen for the column and conditions to be used.
3. Always remove a short length of the column end (approx. 1 cm) **after** introducing the ferrule onto the column. This ensures that no minute particles from the ferrule are contaminating the column.
4. To ensure a clean cut of fused silica tubing always use a high quality tungsten carbide glass knife. This will ensure a clean cut with no jagged edges. **Never try to break tubing by hand.**
5. Ensure that the fittings are secured at the correct positions on the column according to manufacturer's recommendations for column installation.
6. Never overtighten compression fittings.
7. Always test fittings for leaks before use. This can be done using detergent solutions or by a commercial electronic device.

Connections often have to be made between columns, or between a column and **retention gap** (see Chapter 6). Before commercial fittings became available these connections were made by the use of heat-shrinkable PTFE tubing which was an expedient but not very satisfactory method, often producing leaks and absorption effects. Various types of compression fitting are now available, including the **single ferrule connector** and glass taper-fit connectors as shown in Figure 3.7. The single ferrule connector is based on the same principle as the normal compression fitting shown in (a) but only a single ferrule is used for both ends of the two columns to be connected. These ends comprise, in effect, a butt-to-butt connection. A variety of more conventional **two-ferrule** connectors are available commercially where each end of the tubing to be joined is fitted into a stainless steel body similar to the principle illustrated in Figure 3.7a. Glass taper-fit connectors are certainly the simplest devices for fused silica columns and are available from several manufacturers under a variety of trade names. This type of connector is replacing quickly the compression type of fitting for joining fused silica capillaries by virtue of simplicity, and the absence of any reactive metal or polymeric surfaces. The two capillary ends must be cut cleanly as described above and then pushed into the fitting until they seat in the tapered tube. The joint is sealed by the polyimide coating on the outside of the column.

A major advantage of this fitting is that two different sized columns can be connected in this way. Grob has described a number of other possibilities for coupling columns together [2].

Multiway connections are also used when three or more columns need to be connected together. For instance, in dual detector operation the exit flow from the capillary column is split into two separate streams each of which is monitored by a different detector (see Section 3.17). Suitable connectors for a variety of configurations are available can be either of the compression or glass taper-fit type.

3.6 THE INJECTOR

The injector is the device which enables the sample to be applied to the column. This is one of the most critical parts of the system, affecting both the performance of the column and the quality of the analytical results. These features depend very critically on the injector design, its operating conditions and the technique used by the operator. Modern instruments can be fitted with a range of injectors, the choice of which depends on the nature of the sample and the column. Most injectors are heated in a separate unit which is usually attached to the column oven, but can be adjusted independently to the requisite temperature. Usually there are two injector heaters each of which can accommodate at least one, and sometimes two injectors. With **vaporization injectors** the injector temperature is normally set to volatilize all the applied sample very rapidly, i.e. to produce **flash vaporization** conditions. Most commercial instruments will allow injection temperatures up to 350–450 °C. The range of injectors available for most commercial instruments include devices for sample introduction by **bypass, direct, split, split–splitless, on-column,** and **programmed temperature vaporization** methods. These techniques are all described in Chapter 6 together with IUPAC definitions of the terminology. Commercial injectors are interchangeable units that can be fitted as required to the top of the column oven and so a combination of various injection methods can be available on any single instrument.

Nowadays, capillary GC is applied routinely for quality control purposes, and all modern instruments can provide the option of automatic sample introduction. These units were originally called **auto-injectors** but now are normally referred to as **autosamplers**, in common with the nomenclature used for HPLC instrumentation. Autosamplers can be loaded with a number of samples in special vials and a range of injection options can be applied via the software control systems.

3.7 THE CHROMATOGRAPHIC OVEN

As chromatographic retention properties are critically dependent on the column temperature, it is essential to control this variable very accurately and precisely, and to ensure that there are no temperature gradients in the oven space. Any gradients or 'cold spots' will cause a deterioration of chromatographic performance with peak distortion.

Physically, the oven is the largest component of the chromatographic system, and is the central feature to which other components are physically linked. It must be large enough to accommodate the columns required for the intended applications, ranging from capillary to packed, and be provided with mounting facilities which hold the columns in a neat and compact form centrally inside the oven to prevent any possibility of contact between the column and the oven wall. Most GC ovens are heated electrically with forced air circulation by a fan which is usually mounted at the back. The usual temperature range is from ambient to 400–500 °C. Some makes of equipment includes a provision for cooling the oven to below ambient temperatures but the use of **sub-ambient** operation has declined in recent years because many of the separations which previously required low column temperatures can now be performed with thick film, or PLOT columns.

The most important GC oven properties for the potential user are given below.

1. **Minimum temperature**: the minimum temperature at which the oven can run with accurate control. This is usually 10–20 °C above ambient.
2. **Maximum temperature**: the maximum temperature at which the oven can run with accurate control. Values range from 300–500 °C. Columns are now available for high temperature use which can be operated up to 450–500 °C. It is important to note this specification for intended purchases if high temperature operation is required.
3. **Accuracy and precision characteristics**: a modern oven should be controllable to within 0.1 °C of the required temperature and to maintain this value to within 0.1 °C for the duration of the run. Reproducibility of programming conditions from run to run should be within 0.2 °C. In any case the reproducibility of gross retention times should be less than 0.2% relative standard deviation.
4. **Programme rates**: a range of temperature programming rates should be provided, including the ability to change the rate, or to include isothermal periods within the run. Programming rates provided are normally in the range 1 °C min^{-1} up to at least 20 °C min^{-1}.
5. **Rapid cool-down**: this is very important as the cool-down time adds to the overall analysis time. A good oven should be able to cool from 250 to 50 °C, including final stabilization in less than 10 minutes.
6. **Rapid heatup**: on switching on the column oven it should be capable of heating **ballistically** and stabilizing to 300 °C in less than 10 minutes.

3.8 DETECTORS FOR CAPILLARY GC

The function of the detector is to provide a coherent, instantaneous signal in response to the elution of components from the column. Modern capillary GC owes much of its current success to the development of high sensitivity ionization detectors, and in particular to the **flame ionization detector** (FID). This is the most ubiquitous of all the detectors in GC and most laboratory gas chromatographs are equipped with this for general purpose applications.

During the early development of the technique little more was required of the detector than that its response should be linearly related to the amount of each separated component. The precise relationship was derived by calibration, as is the case today, but little attempt was made to use the detector to give information about the nature of the components, or to use known molecular characteristics as a basis for the development of selective detectors. An important feature of the modern technique is the availability of detectors which respond selectively to particular species. IUPAC (see Chapter 1, reference 43) defines a **selective** detector as one which responds to a related group of sample components in the column effluent, and a **specific** detector as one which responds to a single sample component or to a limited number of components having similar chemical characteristics. For instance, the electron capture detector (ECD) is indispensable in the analysis of trace pesticides and other environmentally important compounds and is regarded as a selective detector. Dressler [3] has published a review of selective detectors used in GC. Other widely used detectors are the flame photometric detector (FPD) and the nitrogen phosphorus detector (NPD) and are examples of specific detectors. In contrast a **universal** detector is defined as one which responds to every component in the column effluent except the mobile phase. This latter definition is a little misleading since universal detectors such as the thermal conductivity detector **do**, in fact, respond to the carrier gas and depend for their operation on the **difference** in the response between the carrier gas and the sample components.

The detector signal is normally amplified before being recorded, possibly after further manipulation by a data system. In a basic system the amplified signal is merely monitored by a recorder to produce a chromatogram which is then used for all calculations. This was a normal situation before the advent of integrators and data systems but it was obviously inconvenient and very time-consuming. Although modern data systems are now ubiquitous in many laboratories the chromatogram is still required for visual assessments of separation and the calculation of chromatographic performance such as plate numbers.

3.8.1 Basis of Detector Response

The response of GC detectors depends on the use of appropriate molecular properties and behaviour in the detector environment. This response will be a function either of the solute concentration in the eluting gas stream, or the rate at which it passes through the detector. Detectors which depend on the latter property are called **mass flow detectors**, and are usually based on a destructive mechanism. Those that depend on the concentration of the components in the column effluent are called **concentration detectors** and usually have no destructive effect. Comparing the effects, a mass flow rate response would

CAPILLARY INSTRUMENTATION

normally fall to zero level if the flow rate is turned off, whereas a concentration response would remain at the same level. Thus the two types of detectors are affected differently by carrier gas flow rate with consequential effects on chromatographic sensitivity.

Equation (2.35) gives the mass of solute in the rth plate when a unit mass of solute is applied to the column and the maximum is at the r_mth plate. It follows that if a mass W_i of a single solute is applied to the column of N theoretical plates then the mass in the last plate when the maximum is Δr plates from the exit, is given by,

$$Q_N = \frac{W_i}{\sqrt{2\pi N}} e^{-\left(\frac{\Delta r^2}{2N}\right)} \qquad (3.8)$$

When Q_N is at a maximum $\Delta r = 0$ and the exponential part of this expression must be unity; hence,

$$Q_N(\text{max}) = \frac{W_i}{\sqrt{2\pi N}} \qquad (3.9)$$

Since the retention volume of the solute is V_R, then to elute $Q_N(\text{max})$ from the last plate requires a volume of carrier gas equal to V_R/N. Hence the concentration of the zone at maximum as it elutes from the column is

$$C_{\text{max}} = Q_N(\text{max})N/V_R \qquad (3.10)$$

$$= \frac{W_i}{V_R}\sqrt{\frac{N}{2\pi}} \qquad (3.11)$$

$$= \frac{W_i}{F_{\text{av}}t_R}\sqrt{\frac{N}{2\pi}} \qquad (3.12)$$

Expression (3.11) shows that the maximum concentration of the chromatographic zone is inversely proportional to its retention volume. Since retention volume is unaffected by changes in flow rate then any change in C_{max} will be entirely due to the affect on N. Concentration detectors such as the TCD will therefore produce almost the same peak height if the flow rate is changed, although the peak width, and consequently the area will change.

This effect is illustrated by Figure 3.8a where the second chromatogram is produced at one half of the flow rate used in the first. The effects depicted in the diagram are based on the ideal assumption that the detector sensitivity is not affected by the change in flow rate, but in practice there will be some effect on the overall mechanism of response. Thus, even if the concentration were to

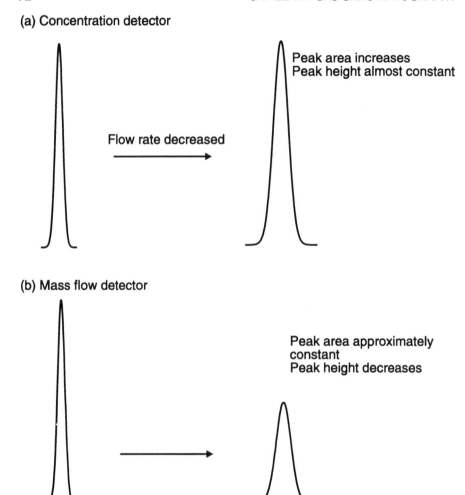

Figure 3.8 Effect of flow rate changes on (a) concentration, and (b) a mass flow detector

remain the same at two different flow rates, a change in sensitivity will cause the response to be different.

It is important not to confuse the effect of flow rate with that of column temperature changes. If the column temperature is decreased then the retention time will also increase as shown in the Figure 3.8, but in this case the change is caused by the effect of temperature on retention volume which would increase rapidly. Consequently the peak height would be lower and the peak area remain about the same.

All the flame detectors used in GC are mass flow sensitive devices and so they give a response which depends on the mass flow rate through the cell, i.e.

$$\text{mass flow rate at maximum (g s}^{-1}) = F_c \times C_{max} \quad (3.13)$$

which, on substituting (3.12) into (3.13)

$$= \frac{W_i}{t_R} \sqrt{\frac{N}{2\pi}} \quad (3.14)$$

The response is inversely proportional to the retention time (neglecting the effect of N) and so any change in the chromatographic conditions which causes a change in retention time will accordingly affect the peak height critically. This effect is illustrated in Figure 3.8b where the flow rate change is the same as for the concentration detector. Now the peak height is also halved and its area remains the same, but modified slightly if the response mechanism changes as noted for concentration detectors. In other words it would be unsafe to assume that response factors used in quantitative analysis remain constant and independent of flow rate changes.

Expression (3.14) shows that the response of any mass flow detector is strongly affected by the carrier gas flow rate. Helium is usually the best carrier gas to use with capillary columns and it is often set at a velocity higher than the optimum. This will also give a gain in detector response as an additional advantage. Wide bore columns (see Chapter 4) which are now used for many applications are normally operated at flow rates approaching those associated with packed columns. The consequent gain in response is a considerable advantage particularly when these columns are used in trace analysis.

3.8.2 Detector Sensitivity

The **sensitivity** of a GC detector is the fundamental relationship between the detector response and the concentration or mass flow rate of solute entering the detector.

For a concentration detector, the sensitivity, S, can be calculated from:

$$S = \frac{A_i F_c}{W_i} = \frac{E}{C_i} \quad (3.15)$$

where A_i is the integrated peak area (in mV min), E is the peak height (in mV), C_i is the concentration of the component in the carrier gas (g cm^{-3}, F_c is the carrier gas volumetric flow rate (ml min^{-1}), and W_i is the mass of the component applied to the column (mg). Thus the product $A_i F_c$ is the peak area measured in terms of **detector voltage** versus **volume** of carrier gas. The dimensions of sensitivity for a concentration detector are therefore mV cm^3 mg^{-1}.

For a mass flow detector the equivalent calculation is,

$$S = \frac{A_i}{W_i} = \frac{E_i}{M_i} \tag{3.16}$$

where A_i is the integrated peak area (A s), E_i is the peak height (A), M_i is the mass rate of substance entering the detector (g s^{-1}). The dimensions of the detector sensitivity is A s g^{-1}.

3.8.3 Background Characteristics

Whatever method of data handling and readout is employed, there are two aspects to the detector output, i.e., background and response. The background is the detector output when operating under chromatographic conditions but when no component is emerging from the column. This output will usually be at a finite level depending on the nature of the detector and the carrier gas, but it will also have a certain level of instability due to factors such as 'bleed' of organic material from the column, electronic noise from the amplifier, and instability arising from the detector itself, perhaps as a result of temperature variations. The background level when no sample is eluting from the column is called the **baseline**, and the background instability is referred to variously as baseline **noise**, **drift**, and **wander**. Although these terms are reasonably self-explanatory, Figure 3.9 illustrates their meanings.

If the instrument is functioning normally then noise and other instability effects should be minimal, although noise will always be evident when the amplifier attenuation is at a low enough setting. The noise level is particularly

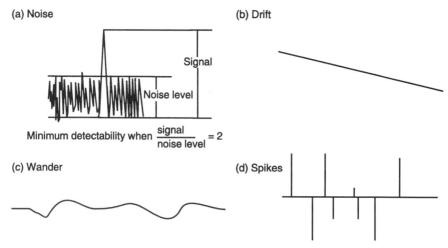

Figure 3.9 Types of baseline instability

important as this limits the amounts of the components that can be detected under the intended analytical conditions. Thus, the **minimum detectable limit** (or **minimum delectability**) is the mass of any stated component, when separated by chromatography under the required analytical conditions, that produces a peak height equal to twice the noise level. The minimum detectable limit should always be determined before attempting trace analysis to see if some form of sample preconcentration is necessary before the analysis. This property depends on the nature of the sample, the baseline stability, the detector, and the chromatographic conditions and it should not be confused with values given by some instrument manufacturers as the **detectivity**. This is a property of the detector alone without any reference to a column or chromatographic conditions.

In many ways the detector can be regarded as the 'window' to the GC, particularly with regard to disturbances in the operation of the instrument or in the quality of the chromatography. Solving these problems is of critical importance to the user and some diagnosis of them is possible from the nature of any baseline irregularities.

3.8.4 Detector Linearity

Another very important property of the detector is its **linearity**. Clearly it is advantageous for the relationship between the detector response and the component concentrations in the sample to be linear, at least over the relevant range. **Response factors** are the ratios of peak areas, or heights, to the mass of each component separated by chromatography, using specified chromatographic conditions. Because it is almost impossible to know with any certainty the absolute masses of the components separated, **relative response factors** to a chosen added standard compound are normally employed. These values are independent of the amount of sample injected provided that the amounts are within the linear range of the detector and are determined by calibration with prepared mixtures of pure compounds (see Chapter 8). These calibration values allow component concentrations to be determined by calculation rather than by graphical interpolation, as would be the case if the response was non-linear.

Detector linearity is usually expressed by the term **linear range**, which according to the IUPAC definition is the range of concentrations (or mass flow) of a substance in the mobile phase at the detector over which the sensitivity of the detector is constant within a specified variation, usually 5%. This detector property should not be confused with the **dynamic range**. The IUPAC definition of dynamic range is the range of concentration or mass flow rates of a substance over which an incremental change in concentration (or mass flow rate) produces an incremental change in detector signal. The dynamic range of the detector is invariably wider than the linear range. This definition of linear range, which includes the maximum allowed deviation, obviates the need to use less well defined term **linear dynamic range**.

The main detector feature are represented in Figure 3.10 from which linear ranges can be determined.

Figure 3.10 is usually plotted on a log–log scale, and for practical purposes other units could be substituted for the two axes, namely, concentration of specific component in the sample for a fixed quantity injected versus peak height or area expressed as an integrator count. In this event the plot would be specific to the particular application, the choice of column, and its operating conditions.

3.8.5 Detector Makeup Flows

As mentioned in Section 3.2, most detectors used in capillary GC have a provision for a separate makeup flow of inert gas. For some detectors this

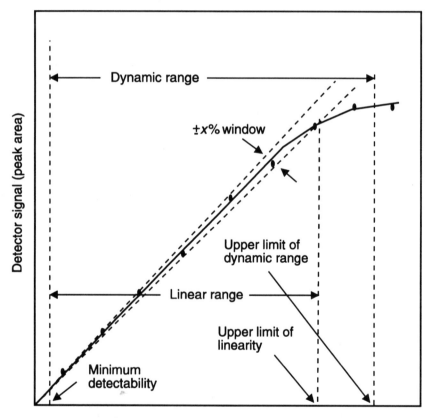

Figure 3.10 Detector response characteristics. (Reproduced with permission from *Pure Appl. Chem.*, **65**, 818. Copyright © 1993 IUPAC)

CAPILLARY INSTRUMENTATION 77

facility is provided to reduced its effective volume, but a further reason may be to improve its response characteristics.

The need or otherwise for this makeup flow is discussed separately with each detector.

3.9 FLAME DETECTORS

The various types of flame detector have been largely responsible for the successful development of capillary columns and have become indispensable to the modern use of the technique. They are all based on combusting the eluted compounds in a hydrogen flame to produce a plasma whose properties are then measured. The first detector of this type, described by Scott [4] was based on the continuous measurement of flame temperature by a thermocouple situated immediately above the hydrogen flame plasma. The temperature varied according to the heats of combustion of the eluted components and the thermocouple registering these variations was monitored by a potentiometer circuit and recorder. Although this detector is no longer used it undoubtedly initiated interest in the use of flame properties for GC detection.

3.10 FLAME IONIZATION DETECTOR (FID)

The FID has become the most widely used detector in gas chromatography following its introduction almost simultaneously by Harley and coworkers [5] and McWilliam and Dewar [6] in the late 1950s. A historical review of the development of the FID was recently published by McWilliam [7]. The appeal of the detector arises from its inherent simplicity, high sensitivity, and incomparable linearity. In spite of these considerable advantages the mechanism is undoubtedly complex and remains incompletely understood. A schematic diagram of the detector is shown in Figure 3.11a and a cross-section of a commercial detector in Figure 3.11b.

The operation of the detector depends on mixing the carrier gas stream after it elutes from the column with sufficient hydrogen to produce a combustible mixture. This mixture is admitted to a capillary jet and burnt in an excess of clean air. The jet must be heat resistant and is usually constructed from stainless steel, platinum, platinum–iridium, and other suitable materials. Some designs include an inner coating of ceramic to reduce the possibility of solute decomposition before it can emerge from the jet. A pair of electrodes in close proximity to the flame, one of which is often the jet itself, is supplied with a polarizing voltage and the resulting plasma current is amplified and measured. Because of the very high resistance of the plasma the amplifier must have a special design for it to accept high impedance

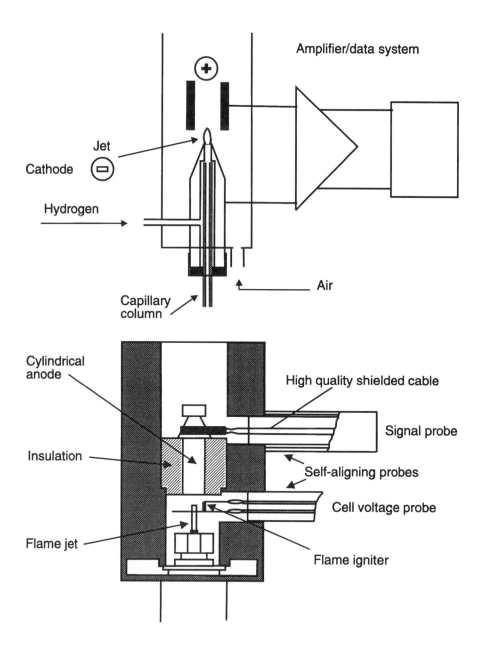

Figure 3.11 (a) Flame ionization detector schematic and (b) cross-section of commercial FID. (Reproduced by permission of Varian Analytical Instruments)

CAPILLARY INSTRUMENTATION

signals and converting them to low impedance outputs for handling by recorders, integrators, and data systems.

3.10.1 Response Mechanism of FID

The FID response mechanism may not be of concern to the analyst as the detector is almost invariably calibrated for quantitative analysis. The detector has always given a very high degree of linearity and sensitivity and improvements in design have not given significant improvements in these properties. It now seems unlikely that a more complete understanding of the response mechanism would change this situation. Nevertheless, studies of detector response characteristics have helped in the prediction of response factors for compounds that are not available in the pure state. For instance Jorgensen and coworkers [8] have described ways of predicting response factors from molecular structure using effective carbon numbers and corrections that need to be applied for the presence of different functional groups. Also, it is well known that compounds of a similar chemical type give similar responses, the major factor being the number of carbon atoms in the molecule.

The main difficulty in any comprehensive study of the FID is the number of interdependent variables, many of which depend on the construction and geometry of the detector. There are some common features, however, related to the rates of formation and recombination of ions, and their probability of reaching the anode to produce a current. The design and operating parameters of all detectors are based on maximizing the probability of ion collection to maximize the sensitivity.

With regard to the FID mechanism, there were some excellent reviews published during the early developmental period, particularly by Boćek and Janák [9], Novák and coworkers [10], and Bolton and McWilliam [11].

According to Ševčik and coworkers [12], the FID mechanism is based on two major stages, namely,

1. radical formation requiring the absence of oxygen, e.g.

$$CH_3CH_2CH_2CH_3 \rightarrow 2CH_3^{\cdot} + 2CH_2^{\cdot} \qquad (3.16a)$$

or

$$C_6H_6 \rightarrow 6CH^{\cdot} \qquad (3.16b)$$

2. chemical ionization by excited oxygen states, i.e.,

$$\begin{aligned} CH_3^{\cdot} + O^{\cdot} &\rightarrow CH_2O^{\cdot} + H^+ \\ CH_2O^{\cdot} + O^{\cdot} &\rightarrow CHO^+ + OH^- \\ CH_3^{\cdot} + O_2^{\cdot} &\rightarrow CH_2O^{\cdot} + OH^- \\ CH_2O^{\cdot} + O^{\cdot} &\rightarrow CHO^+ + OH^- \end{aligned} \qquad (3.16c)$$

These proposed mechanisms are speculative but they do give rise to some interesting predictions which are supported in practice. For instance, if water is present in the carrier gas the primary $CHO+$ ion formed in the secondary stage of the process reacts rapidly with the water to form a hydrated proton, as follows:

$$CHO^+ + H_2O \rightarrow H_3O^+ + CO \tag{3.16d}$$

This reaction increases the probability of ion recombination, hence reducing response. Such quenching of FID response has indeed been noticed when excessive amounts of water are present in the carrier gas [13]. However, this should not deter users from performing separations of aqueous samples with the FID. Aqueous analyses are carried out frequently and accurately by capillary GC, particularly for environmental applications. Probably the relatively small quantities of water in the sample compared with the amounts of carrier gas and fuel gases minimizes the likelihood of quenching effects. All suggested FID mechanisms are based on the presence of carbon in the solute molecule, and experience shows that this is the main factor governing the detector response. Compounds which do not contain carbon such as permanent gases and water give no response, and the presence of heteroatoms such as nitrogen, oxygen and sulphur, tend to reduce the detector response as in compounds such as CS_2 and formic acid. The lack of detector response to water is particularly useful for the analysis of aqueous samples as a large solvent peak is thereby avoided which would otherwise obscure early components.

When only carrier gas is eluting from the column there is no possibility of ionization and so the low background current enables a high degree of amplification thereby ensuring a high sensitivity. The ionization current is associated with the efficiency of ion formation and collection which in turn depends on the detector design and its operating conditions. These conditions include the flow rates of the various gases and the polarization voltage. Manufacturers usually employ a preset voltage for the latter in the saturation region, i.e., such that any increase in potential difference does not produce an increase in the current. A variety of voltages are used in instruments, the precise value seemingly of little importance, but typically in the range 100–300 volts.

3.10.2 Effect of Supply Gases on FID Response

Three auxiliary gas supplies are required in addition to the carrier gas, namely, hydrogen for the fuel gas, air to support the combustion and inert makeup gas. The hydrogen supply must be pure and free from organic impurities to avoid unwanted background and noise. It can be supplied either from a cylinder or an electrolytic hydrogen generator. A charcoal filter should be included in the hydrogen supply line to remove any organic contaminants.

CAPILLARY INSTRUMENTATION

The air supply should also be free from organic contaminants for the same reason, and again this should be derived either from an air cylinder or generator. As mentioned earlier in this chapter, it is not advisable to employ atmospheric air pumped into the system by compressors, because of undesirable pulsation effects and the probable contamination from this source.

The response of the FID depends on the individual flow rates of the three gases, and it is essential that the operator adjusts the various flow rates to the appropriate values for the detector. These values will depend on the particular design of the detector and so it is vital that manufacturer's recommendations be followed.

Figure 3.12 illustrates the general effect of increasing the air supply to the detector.

When there is sufficient air to support combustion the detector response increases rapidly to reach a plateau region, usually at about 200–300 ml min^{-1}. Higher flow rates beyond the plateau region cause the flame to become

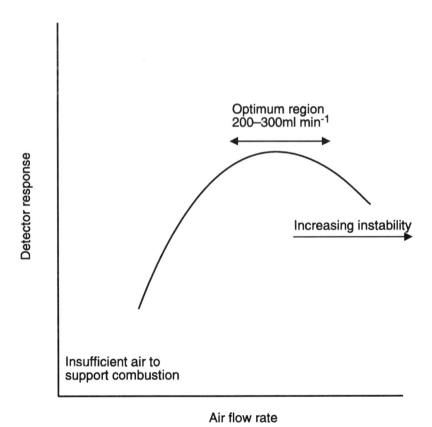

Figure 3.12 FID response characteristics versus air flow rate

unstable, producing increasing noise with a declining response and so eventually the flame will be extinguished. If the detector is operated with low air flow rates before the plateau region then the sensitivity will be lower than optimum and the detector response will become very sensitive to minor fluctuations in the air supply. Also, the flame is likely to be extinguished when large solvent zones elute from the column. The correct operating flow rate is, of course, in the plateau region and the manufacturer's recommendations should always be adopted.

Assuming that the detector is operated at the correct air flow rate then the effects of the hydrogen and carrier gas flow rates can be represented by a response surface similar to that shown in Figure 3.13.

The region of maximum response requires the presence of a significant flow rate of inert gas. Capillary columns are operated at carrier gas flow rates which are much lower than this and so a makeup flow of carrier gas is introduced as discussed previously in Section 3.8.5 serving in this case to optimize the detector response. This makeup flow has other beneficial effects, i.e. reducing the flame temperature hence stabilizing the detector, prolonging the lifetime of the jet, and purging any unswept volumes.

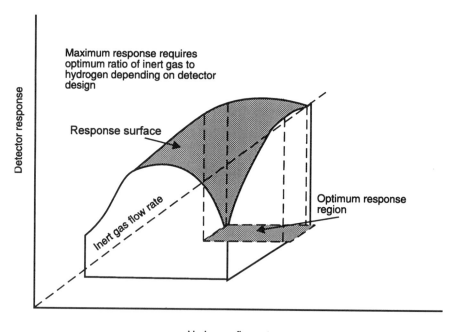

Figure 3.13 Effects of hydrogen and inert gas flow rates on FID response characteristics

CAPILLARY INSTRUMENTATION

Another important operating feature is the detector temperature. If the detector was not heated then water formed during combustion of the hydrogen would condense inside the detector causing the flame to be extinguished. The main reason for heating the detector, however, is to avoid any condensation of the sample components as they emerge from the column as there is a considerable increase in their concentration when they vacate the stationary phase. Thus some condensation could occur in an unheated detector causing serious baseline instability and badly misshapen peaks. Flame detectors can normally be heated to about 400 °C although the precise temperature is not critical. It should be selected to be at about the mid-boiling point of the sample, i.e. sufficiently high to avoid condensation but not so unnecessarily high that sample decomposition might occur.

Flame ionization detectors do not give many operational problems provided that they are properly maintained and used optimally. The detector jet naturally gets very hot and can become contaminated or even blocked with decomposition products such as carbon, silica from polysiloxane phases or derivatization reagents, and particulate matter.

If the jet becomes irreparably damaged it can be replaced, and spares should always be available. Most commercial flame detectors are easy to dismantle to allow access for servicing and cleaning. This should be done on a regular basis as it is always better to avoid problems than to allow them to develop by neglect.

Another sensitive detector component is the anode and the signal cable. This must be a very high quality connection with no possibility of leakage between the conductors. Any damage or vibration to this lead will cause detector response problems and baseline noise.

3.11 THE NITROGEN PHOSPHORUS DETECTOR (NPD)

The NPD is referred to by several alternative names: the **alkali flame** detector (AFD), the **flame thermionic** detector (FTD) or **thermionic specific** detector (TSD), etc. The modern detector evolved during the period 1960–1970 at a time when extensive efforts were being applied to the development of new selective detectors.

Guiffrida [14] first noted that the presence of an alkali metal salt in the flame of an FID caused an enhancement of its response towards organophosphorus and halogen compounds. However it is for the selective response towards **phosphorus** and **nitrogen** compounds that the NPD is now mainly designed. Numerous jet and electrode configurations have been used, with a variety of operating conditions but the most troublesome aspect has been the nature of the alkali metal salt. Early detectors consumed this rapidly and so it had to be replaced frequently. The modern detector employs a rubidium or caesium salt incorporated in a special, separately heated refractory bead.

Figure 3.14 shows a cross-sectional diagram of a modern commercial detector of this type.

In this design the column effluent is combined with a very low hydrogen flow (2–5 ml min^{-1}) which is admitted through the detector base. This is insufficient to produce a self-sustaining flame but gives a low temperature plasma surrounding the bead which is electrically heated to 600–800 °C. The bead has a glass–ceramic composition incorporating, in this case, a caesium salt. The NPD is electronically similar to the FID, with a cylindrical anode at a positive potential with respect to the jet and bead. In contrast to the FID the NPD does not require a separate makeup flow although this may be provided in some instruments.

The NPD is highly specific device, used widely for trace analysis. Typical applications are the determination of trace levels of pesticides in environmental samples, and the toxicological examination of clinical samples for nitrogen drugs and their metabolites. Figure 3.15a shows a typical application; the analysis of trace amounts of nitrosamines in a herbicide.

3.11.1 Mechanism of NPD Response

In the design shown in Figure 3.14, the flame plasma causes partial combustion of the eluted components with the formation of semistable intermediate radicles, as with the FID. If nitrogen or phosphorus compounds are present then these interact with ions formed by the heated bead to produce ions which

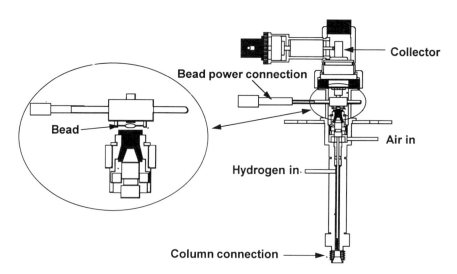

Figure 3.14 Schematic diagram of nitrogen phosphorus detector (NPD). (Reproduced by permission of Perkin Elmer Corporation)

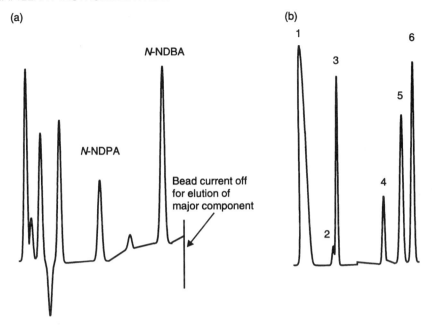

Figure 3.15 (a) Trace analysis of nitrosamines in herbicides using a nitrogen phosphorus detector and (b) low ppm level sulphur compounds in natural gas using a flame photometric detector. Key: 1 = methane/ethane; 2 = propane; 3 = carbonyl sulphide; 4 = carbon disulphide; 5 = *tert*-butyl mercaptan; 6 = dimethyl sulphide. (Reproduced by permission of Varian Analytical Instruments)

are in turn collected by the anode. Although there have been many attempts to elucidate the mechanism of the NPD and the reasons for its extreme specificity, it remains even less understood than the FID mechanism [15–17]. The main controversy is the exact nature of the function of the heated bead, i.e., whether this causes the formation of excited alkali metal atoms which interact with short lived N or P containing radicles in the plasma as a homogeneous gas phase reaction, or whether the interaction is a surface phenomenon. Even greater controversy has been introduced by van de Weijer and coworkers following emission and laser-induced fluorescence measurements. These authors suggest that the mechanism of the NPD response involves sodium ions rather than the dosed alkali metal, and that the bead would operate equally well if simply made of soda glass [18].

For the average user, the important properties of the NPD are its specificity towards nitrogen and phosphorus compounds, its linearity, and its general level of sensitivity. These features are included in Table 3.3 at the end of this chapter which gives a comparison of the properties of the main GC detectors.

3.12 FLAME PHOTOMETRIC DETECTOR (FPD)

The author can claim to be the first to describe the use of the emissive properties of a flame as a detector for GC. This was probably the first attempt to employ a GC detector in a selective sense, in this case, to differentiate between aromatic and aliphatic compounds [19].

It should be possible to detect many heteroatoms, including metallic elements by this means, and in fact photoemissive properties have been utilized in GC for the selective detection of elements such as halogens, nitrogen, boron, selenium, etc. [21]. The measurement of photoemission at selected wavelengths is a very attractive way to achieve selectivity, as it involves more clearly understood principles than other flame detectors. Although the FPD is a standard commercial option for most instruments its use is now almost exclusively limited to the selective detection of **phosphorus** and **sulphur containing** compounds. Clearly there is still considerable scope for widening the range of application.

The modern FPD is based on premixing the eluent carrier gas with air and burning this in a hydrogen atmosphere. Several commercial detectors use a dual stacked jet arrangement giving two vertically positioned flames. An example of this type of design is shown in Figure 3.16.

The lower flame produces highly reduced species from all the organic material eluting from the column, many of which emit at wavelengths that

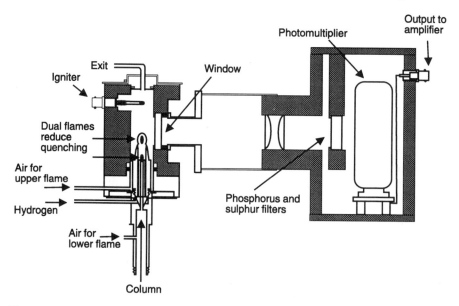

Figure 3.16 Schematic diagram of a flame photometric detector. (Reproduced by permission of Varian Analytical Instruments)

would either obscure or quench the required emissions from the P or N compounds. The second flame is an optimized flame which excites P or N containing radicles to HPO^* and S_2^* species respectively. These have very characteristic emission spectra giving maximum emission intensities at wavelengths of 526 nm for phosphorus and 394 nm for sulphur. The emitted radiation is monitored by the photomultiplier, and the resulting current amplified. As with the NPD, this detector does not normally require a separate makeup flow as the jet volume is rapidly purged by the air supply which is premixed with the carrier gas.

The sensitivity of the detector is a little higher for phosphorus compounds than for sulphur compounds (see Table 3.3), but a significant drawback in the sulphur mode is its non-linearity. This is due to the formation of the S_2 radicle upon which the selective response is based and so the detector output is theoretically proportional to the **square** of the sulphur mass flow rate. Commercial data systems for the detector are often equipped with appropriate software which provides a square root linearizer function. This should be used with care, however, and checked for linearity because considerable deviations from the square law are observed in practice. The general relationship can be stated as:

$$R = AS^n \qquad (3.17)$$

where A = constant, S = mass flow rate of sulphur-containing compound and n = constant with a practical value of between 1.5 and 2.5.

The detector is used widely for the trace determination of sulphur compounds, particularly when they are in the presence of much larger amounts of other types of compounds such as hydrocarbons, etc. A typical application showing the presence of sulphur odorants in natural gas is given in Figure 3.15b.

3.13 THE ELECTRON CAPTURE DETECTOR (ECD)

The modern electron capture detector is the next most important detector in use today after the FID. It is a highly selective detector dependent on the **electron affinity** of the molecule. This is a very difficult property to predict except in very general terms, but there are many types of molecule which can produce high responses. The ECD is used mostly for the trace measurement of **halogen compounds** particularly in environmental applications involving insecticide and herbicide residues.

The detector was introduced by Lovelock [22] following earlier work on a series of ionization detectors. The detector consists of a simple flow cell containing a beta-emitting radioactive source. Two electrodes inside the cell create a low energy polarizing field by the application of a low voltage. As with other ionization detectors there are numerous interactions that can occur within the cell and discussions of these are well documented in the literature. A survey published by Zlatkis and Poole gives an overall view of the theory and practice of the detector [23]. A more recent study by Chen and coworkers have presented a kinetic model to explain the ECD mechanism [24].

3.13.1 ECD Mechanism

A basic explanation of the ECD response is as follows. When only pure carrier gas is passing through the cell a constant current is produced as a result of some ionization of the gas by β electrons from the radioactive source and ion discharge at the polarized electrodes. Thus, there is continuing presence of free electrons and positive ions from the carrier gas. If the polarizing potential were to be gradually increased from zero then this background current would also increase but eventually it would reach a constant saturation value (I_o) when all the electrons present in the cell are discharged at the anode. If a constant concentration of an electron absorbing vapour is now introduced into the cell a number of reactions can take place, but the principle **required** reaction is,

$$AB + e^- \rightarrow AB^- \tag{3.17a}$$

The negative ions have a lower mobility than free electrons and so are rapidly scavenged by the positive ions present in the cell before they can reach the anode. Since the rate constant for the recombination of oppositely charged ions is approximately 10^6 times faster than that between positive ions and electrons there will be a nett **decrease** in the electron population, and consequently a **decrease** in the saturation current to I A. The relationship between the saturation current I_o and I should ideally follow the exponential law:

$$I = I_0 e^{-K_a c} \tag{3.18}$$

where c = concentration of electron-capturing vapour in the detector cell and K_a is a constant depending on the electron affinity of the compound. The detector is therefore **concentration** sensitive, although in practice this is largely irrelevant as changes in flow rate through the cell affects the constant in (3.18) due to changes in electron capture efficiency.

If the detector were to be operated in this rather simple way it would, according to (3.18), produce a non-linear response. In addition, many interfering reactions are possible which obscure the basic electron-capturing process, including the formation of space charges in the region of the anode and actual ionization of the sample molecules. Ionization, of course, would tend to increase the cell current, opposing the effect of electron capture causing peak distortion and resulting in negative, or partly reversed peaks in the chromatogram.

3.13.2 Practical Operating Features of the ECD

In the modern detector the beta-emitting radioisotope ^{63}Ni is commonly used as an electron source because of its low activity and high temperature stability to above 400 °C. With regard to the operation of the detector, problems such as sample ionization, and space charge effects are reduced by using the **pulse modulated current feedback mode**. This involves the application of short duration square wave voltage pulses to the electrodes so that the cell is under zero field conditions for most of the time. The time interval between the pulses is varied by feedback so as to maintain a constant average current in the cell, and this improves the sensitivity and linearity by several orders of magnitude. The pulse mode of operation serves to reduce the free electron energy to thermal levels, hence avoiding the possibility of ionization. Another important feature is the choice of carrier gas. Obviously this must be pure and preferably free from any moisture, but the normal gases for capillary operation are acceptable. However, as with the FID, a provision for a makeup flow is usually provided, but in this case it serves a dual purpose. When the ECD is used with a capillary column, it is nearly always necessary to reduce its effective volume with makeup flow. Although this can be accomplished with carrier gas, it is preferable to use a special gas consisting of argon containing 5–10% methane which has the effect of reducing the electron energy in the cell by non-elastic collision. The presence of methane is necessary to inhibit the formation of excited meta-stable states when using argon; otherwise, excited argon molecules can cause ionization of the eluting compounds by energy transfer.

A sectional diagram of a commercial detector is shown in Figure 3.17.

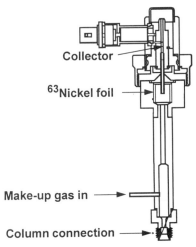

Figure 3.17 Schematic diagram of the electron capture detector. (Reproduced by permission of Perkin Elmer Corporation)

Peak identification:
1. α-HCH
2. HCB
3. β-HCH
4. γ-HCH
5. aldrin
6. o.p'-DDE
7. α-endosulphan
8. p.p'-DDE
9. dieldrin
10. o.p'-DDD
11. endrin
12. β-endosulphan
13. p.p'-DDD
14. o.p'-DDT
15. p.p'-DDT

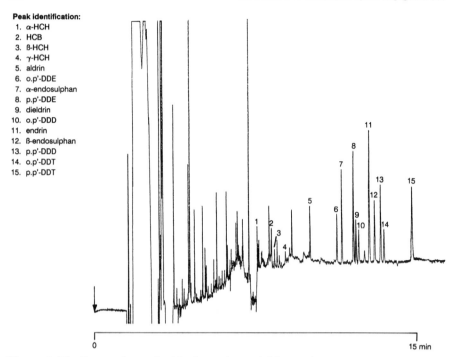

Figure 3.18 Separation of chlorinated pesticides using an electron capture detector. Column: 50 m × 0.25 mm (film thickness = 0.12 μm) CPsil8CB for pesticides; temperature 65 to 225 at 35 °C min^{-1}. H$_2$ carrier gas; on-column injection, 0.5 μ; detector ECD; sample concentrations 2 pg ml^{-1}. (Reproduced with permission of Chrompack International BV)

The radioactive source in this design consists of a 15 mCi (1 Ci = 3.7 × 10^{10} Bq) ^{63}Ni isotope foil. The cell is operated as described above, in the constant current pulse mode, and this is electronically amplified and digitized. An typical example of the use of the ECD for the ultratrace analysis of recovered chlorinated pesticides is given in Figure 3.18.

The response of the ECD depends on the electron affinity of the molecule, and this can vary enormously, even for compounds of a similar type, as indicated in Table 3.2. Careful calibration is essential.

The ECD response for some halogenated hydrocarbons is often disappointingly low and a technique known as **selective electron capture sensitization** [21] can sometimes improve the response by as much by several orders of magnitude. This technique involves dosing the detector via the makeup flow with a minute concentration of an electron-capturing gas such as oxygen, or nitric oxide. Negative molecular ions such as O$_2^-$ and NO$^-$ are formed in the detector as a primary reaction and these subsequently can interact with compounds of low electron affinity to form stable negative ions. Since the nett

CAPILLARY INSTRUMENTATION

Table 3.2 Relative response of the ECD to organic compounds
Reproduced by permission of Elsevier Science, from C.F. and S.K. Poale (1991)
Chromatography Today, p. 278.

General organic compounds		Halocarbons	
Compound	Relative response	Compound	Relative response
Benzene	0.06	$CF_3CF_2CF_3$	1
Acetone	0.5	CF_3Cl	3.3
Di-*n*-butyl ether	0.6	$CF_2=CFCl$	100
Methyl butyrate	0.9	CF_3CF_2cl	170
1-Butanol	1	$CF_2=CCl_2$	670
1-Chlorobutane	1	CF_2Cl_2	3×10^4
1,4-dichlorobutane	15	$CHCl_3$	3.3×10^4
Chlorobenzene	75	$CHCl=CCl_2$	6.7×10^4
1,1-Dichlorobutane	111	CF_3Br	8.7×10^4
1-Bromobutane	280	$CF_2ClCFCl_2$	1.6×10^5
Bromobenzene	450	$CF_3CHClBr$	4×10^5
Chloroform	6×10^4	$CF_3CF_3CF_2I$	6×10^5
1-Iodobutane	9×10^4	CF_2BrCF_2Br	7.7×10^5
Carbon tetrachloride	4×10^5	$CfCl_3$	1.2×10^5

process consumes an electron, the detector response mechanism is similar to that of the basic ECD, but with considerable enhancement.

3.14 THE THERMAL CONDUCTIVITY DETECTOR (TCD)

The TCD, formally known as the **katharometer**, was probably the first detector to be used for gas chromatography, mainly in connection with earlier forms of gas adsorption chromatography. It quickly became the most popular detector for gas–liquid chromatography in the period 1952–1958, its main attraction being its simplicity and universality of response. This situation began to change on the advent of the FID and other forms of ionization detector because of their much higher sensitivity and greater compatibility with the capillary column.

3.14.1 Mechanism of TCD Operation

A typical configuration for the TCD is shown in Figure 3.19, although in practice the detector is available in a variety of geometries and configurations. The detector is usually constructed from a solid block of metal such as stainless steel, brass, copper, etc., and held at a constant temperature, as with the other detectors described in this chapter.

In this design there are two pairs of flow cells, each pair receiving a separate

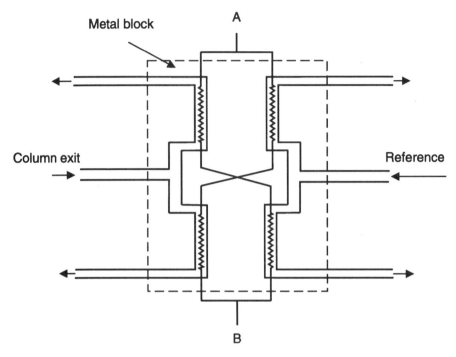

Figure 3.19 Typical thermal conductivity detector bridge circuitry

flow of carrier gas. One of these is the eluent flow from the column and the other is a reference flow of pure carrier gas. The reference cells may simply be filled with pure gas, or connected to an identical column which is supplied with pure carrier gas only. Each of the four cells contains a filament of fine wire comprising metals such as platinum, tungsten, nickel, etc., connected as a Wheatstone bridge network.

A potential difference is applied between points A and B to supply the bridge with current causing the filaments to heat up. This in turn changes their resistance by an amount which depends on their temperature coefficients of resistance. The final filament temperature will depend on the cell block temperature, the supplied voltage, and the rate at which heat is conducted away from the filament, i.e. the thermal conductivity of the gas. If both gas streams are identical, as is the case when only pure carrier gas is present, then all four filaments will be at the same temperature, and the bridge will be in balance. Any change in the thermal conductivity of the sample gas, however, will change the filament temperature in one pair of cells only because of the change in the rate at which heat is conducted away from the relevant filaments. In this design the four filaments are connected so that each gas stream passes through cells connected as **opposite** arms of the bridge. This produces a resultant out-of-balance potential which is recorded to give the chromatogram.

CAPILLARY INSTRUMENTATION

For low concentrations of solute in the carrier gas, the thermal conductivity of the gas is proportional to the concentration of the solute, and so the TCD is a concentration sensitive detector in contrast to the flame detectors discussed previously.

This **constant voltage** mode of operation gave a satisfactory performance with early packed columns but the sensitivity and linear range under these conditions would be inadequate for capillary column operation. Another severe limitation was the relatively large volume of the flow cells which according to (3.3) would cause a severe loss of column plate number. The effective cell volume can be reduced by using a makeup flow but this would also reduce the detector sensitivity because of the resulting dilution effect. Modern detectors can now be manufactured with very low cell volumes. For instance the detector used in the micro-GC shown in Figure 3.2 is made by an etching process which gives a cell volume of only a few nanolitres, a photograph of which is shown in Figure 3.20.

Commercial TCDs for capillary operation, are called micro-TCDs, and will usually have a cell volume of 10–20 μl. The mode of operation of the modern device has also progressed significantly. Detectors are now operated using either **constant current** or **constant filament temperature** circuits, thereby producing much higher sensitivities and wider linear ranges.

3.14.2 Choice of Carrier Gas for the TCD

The sensitivity of the TCD depends on the difference in thermal conductivity between the carrier gas and the sample components, and so the choice of

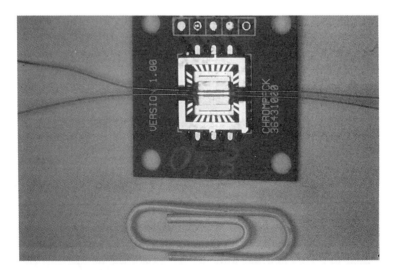

Figure 3.20 Nano-volume TCD as used in micro-GC. (Reproduced with permission of Chrompack International BV)

carrier gas is critical to satisfactory performance. In general, the thermal conductivities of gases are related to their molecular weights and so low molecular weight gases have higher thermal conductivities than heavy ones. Nitrogen and argon have similar conductivities to many organic vapours and consequently are unsuitable for use with the TCD, except to measure hydrogen or helium. Using either of these as carrier gases would give both positive and negative peaks depending on whether the thermal conductivities of the components were greater or less than that of the carrier gas. Hydrogen and helium are the lightest possible gases and have much higher thermal conductivities than any other gas and so either of these will give a satisfactory performance. Fortunately, this recommendation is in agreement with the best choice of carrier gas for most capillary columns.

As mentioned earlier in this chapter, the TCD is one of the few **universal** detectors available and is once again becoming popular for certain applications, principally for the measurement of permanent gases and water to which the FID does not respond. Analysis of this type is now commonly performed with capillary columns coated with adsorbent materials (namely, PLOT columns in Chapter 5) and their use necessarily involves the use of microversions of the detector. An example of this is shown in Figure 3.21 for a PLOT column coated with a porous polymer.

3.14.3 TCD Response Characteristics

Although the response mechanism of the TCD is reasonably well understood it would be unsafe in practice to use thermal conductivity data to calculate response factors, even if such data were available. There are several possible disturbing factors which can affect the response, such as cell geometry, flow rate and cell temperature.

Although the TCD is a concentration detector whose response should be unaffected by changes in carrier gas flow rate, the practical detector will be affected. This is due to the removal of some heat from the cell by conduction, which in turn depends on the heat capacity of the gas. The faster the gas flow rate the more heat will be conducted away from the cell and this can cause a change in detector sensitivity, even producing non-linearity in extreme cases.

Another effect sometimes seen with heavy carrier gases is partial peak reversal due to the competing nature of thermal conductivity and heat capacity. This effect is not usually a problem, however, using hydrogen or helium carrier gas.

Various flow geometries have been used by manufacturers to reduce this flow dependency, although possibly because of the improvements in the mode of operation, etc., most microvolume detectors for capillary operation seem to have reverted to a simple direct flow design. Also, the number of flow cells used in the detector can vary from four, as shown in Figure 3.19, to a single

CAPILLARY INSTRUMENTATION

Peak identification

1. helium
2. carbon monoxide
3. carbon dioxide
4. nitrous oxide (N₂O)
5. hydrogen sulfide

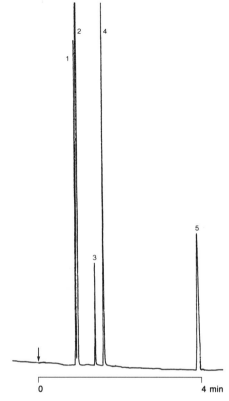

Figure 3.21 Detection of gases by the thermal conductivity detector. Column: 25 m × 0.32 mm fused silica Poraplot Q (porous polymer PLOT). Temperature: 25 °C; split injection; detector: micro-TCD; hydrogen carrier gas. (Reproduced by permission of Chrompack International BV)

cell connected to the column exit, with all compensation carried out electronically. A single cell design simplifies the detector operation from the user's point of view, but obviously the electronic compensation must be very accurate and precise.

Equation (3.12) shows that the maximum concentration of any component eluting from the column, and hence the peak height it gives with the TCD, depends on the mass of component applied to the column, its retention volume and the theoretical plate number. However, the retention volume is roughly proportional to the amount of stationary phase in the column cross-section for equal column lengths. If W_{max} is the **maximum** mass of any solute that can be applied to the column before any overloading effects show up then W_{max}/V_r will remain approximately constant. Thus, if we compare two different columns, e.g. a packed and a capillary column, both loaded with their respective maximum amounts of sample, then the peak heights obtained from the two columns will be about the same provided that the respective detectors are compatible with the two columns and have equivalent sensitivities. The TCD therefore retains about the same chromatographic sensitivity regardless of the type of column; this is a major difference to the behaviour of mass flow sensitive detectors which give a declining response as the column is scaled down. This property is typical of concentration detectors in general and is a considerable advantage when they are employed with small bore columns, with their much lower capacity.

3.14.4 Factors Affecting Stability and Response of the TCD

One of the most important factors in the operation of the TCD is the effect of temperature which must be controlled much more accurately than is the case with flame detectors. Any variation of the detector temperature will cause baseline instability.

The presence of the very hot wire filament in the eluting gas stream can cause thermal or catalytic decomposition of labile compounds. Many sulphur compounds are prone to this problem and usually this causes an irreversible baseline shift after the compound has emerged. Similarly some unsaturated compounds may undergo hydrogenation when they come into contact with the filament if hydrogen is used as carrier.

The response of the detector is obviously very dependent on the filament current thereby necessitating the use of very stable power supplies. Also, it is always necessary to ensure that the carrier gas is turned on and is passing through the detector before the power is switched on, otherwise the filaments are likely to burn out.

3.15 THE PHOTOIONIZATION DETECTOR (PID)

The principle of photoionization was first used by Lovelock as a GC detection method [25]. In the earlier form of the detector the column eluent entered a high vacuum chamber through a hollow anode. The chamber was provided with two electrodes at a sufficiently high potential difference to produce a glow discharge, and photons from the discharge had sufficient energy to ionize organic compounds present in the carrier gas.

This method of detection did not become popular because of the high vacuum at the end of the column, but later detectors were designed in which the detector cell was separated from the high vacuum energy source as shown diagrammatically in Figure 3.22.

The microcell is now operated near to atmospheric pressure and the energy source is a UV lamp filled with xenon, krypton or argon, depending on the

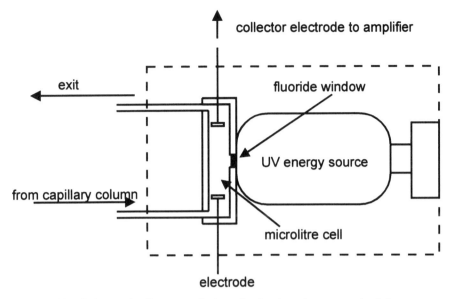

Figure 3.22 Schematic diagram of photoionization detector principle

ionization potential of the sample components. The principle is simple, depending on energization of the molecule by photons from the energy source, followed by subsequent ionization of the energized molecule and amplification of the resulting current between the polarizing electrodes. All compounds having ionization potentials below the energy output from the UV source will give a response and a wide linear range.

The detector is non-destructive and considerably more sensitive than the FID for many compounds, particularly substituted aromatics and cyclic compounds. Response is poor towards permanent gases and low molecular weight alkanes. Langhorst has compiled response data for a number of compounds for the PID [26].

An important advantage of the PID is its simplicity and the lack of any need for auxiliary gases except perhaps makeup flow when using capillary columns. It is therefore useful in hazardous situations where the use of flame detectors would be prohibitive.

3.16 CHEMILUMINESCENCE DETECTORS (CLD)

Chemiluminescence describes the emission of photons by excited chemical species that have been produced by appropriate chemical reactions. Since there are relatively few possibilities for this process to occur then the potential for highly specific GC detectors becomes obvious. The most important applications for CL detectors are for nitrogen or sulphur-containing compounds in concentrations as low as the picogram level.

CL detectors are normally based on a prior gas phase reaction of the solutes to form nitric oxide, in the case of nitrogen compounds, and the sulphur monoxide free radical in the case of sulphur compounds. These then react with ozone to form the excited species which subsequently releases photons which are detected by a photomultiplier. The prior gas phase reaction for nitrogen compounds is usually one of pyrolysis or bond cleavage carried out thermally. This reaction is shown as follows for nitrosamines, one of the important applications:

$$NR_2 - N = O \rightarrow \cdot NO + \cdot NR_2$$
$$\cdot NO + O_3 \rightarrow NO_2 \cdot \rightarrow NO_2 + h\nu \quad (3.19)$$

Although there are several commercial detectors which use variations of the cleavage and reaction conditions, one method is to carry out the prior cleavage reaction catalytically at low temperature (namely <300 °C). The eluting gas is then passed through a cold trap to condense interfering organic volatiles, and then, through a low volume restrictor, into an evacuated chamber containing a slow bleed of ozone where the excitation and emission reactions occur.

For specific sulphur detection the initial reaction to form the SO radical is carried out by combustion in a hydrogen diffusion flame. Again, as for nitrogen detection, the secondary process involves reaction with oxone in an evacuated chamber.

$$RS + O_2 \rightarrow SO + \text{other reaction products}$$
$$SO + O_3 \rightarrow SO_2 * + O_2 \quad (3.20)$$
$$SO_2 * \rightarrow SO_2 + h\nu$$

Sulphur detection by chemiluminescence has a considerable advantage over the use of the FPD in that it is intrinsically linear, with a linear response range of $>10^3$ and detection limits of $10^{-12} - 10^{-13}$ g sulphur per second.

3.17 DUAL DETECTOR OPERATION

A technique that can be used with almost any general purpose capillary GC is to split the capillary column eluent into two streams to feed two different detectors, typically an FID and a selective detector such as the ECD or NPD. This double output gives more detailed information about the composition of the sample, and the distribution of specific types of compound in relation to the overall sample. This type of application is less popular today because of the facilities available with data systems which enable stored chromatograms to be superimposed and examined in a far more versatile manner.

3.18 CAPILLARY GC COUPLED WITH MASS SPECTROSCOPY (GCMS)

The combination of the separation power of capillary GC with the identification abilities of mass spectroscopy provides probably the most powerful separation–identification technique available today. Capillary GC, by itself, can offer strong evidence of identity from retention characteristics, particularly if the source of the sample is known, but this evidence is **never** conclusive and would never be accepted as a legal proof of identity. Such situations are commonplace in environmental monitoring where samples are frequently complex. It is essential that levels of toxic compounds in atmospheric emissions, aqueous discharges, etc., are kept below specified limits, but these will often coelute with other compounds when subjected to capillary GC. The use of selective or specific detectors which only respond to the compounds of interest can sometimes help but the most acceptable option is to apply GCMS.

Mass spectrometry was a well-established technique long before the advent of GC, but has always involved complex and expensive instrumentation. It is not surprising, therefore, that in coupling the two techniques together considerable progress has been made in the development of economical GCMS combinations using several techniques in the sequential processes of ionization, ion focusing and detection.

3.18.1 Ionization Processes

In the traditional mass spectrometer, ionization of sample molecules is carried out in a high vacuum [typically $<10^{-5}$ Torr (1 Torr = 1 mmHg $\approx$ 133 Pa)] by direct **electron impact**, usually at an electron energy of 70 eV, and this produces a range of ions which are characteristic of the molecule. Most of the ions formed in this way are positive parent ions, a few multiple charged ions, and some negative ones. The ions are focused into a narrow beam and then subsequently separated according to their mass to charge (m/z) ratios.

Electron impact ionization is a high energy form of fragmentation in which many of the primary ions may be additionally fragmented and important molecular information subsequently lost. In such circumstances **chemical ionization** (CI) may be preferred which depends on introducing a reagent gas into the ion source to form an ionic plasma at relatively higher pressures (up to 2 Torr) than that required for electron impact. This produces a soft ionization condition based on a secondary reaction between reagent ions and sample molecules to give stable molecular ions without excessive additional fragmentation. The most common reagent gases for CI mass spectrometry are methane and its higher homologues, but there is considerable scope for using a variety of other reagent gases. Thus a simple description of the ionization process where methane is used can be described as follows:

$$CH_4 + e^- \longrightarrow CH_3^- + H^+$$
$$CH_4 + H^+ \longrightarrow CH_5^+$$

(plasma formation by electron impact with excess reagent gas)

$$M + CH_5^+ \longrightarrow MH^+ + CH_4$$

(chemical ionization where M is the sample molecule)

In practice, a number of reagent ions would be formed in the primary process but the basic principle is the same, i.e., that of relative simple soft ionization of the sample molecules.

3.18.2 Ion Focusing

The character of the relative ion abundance produced following ionization is achieved by focusing, or selecting the different species according to their m/z ratios, and the ability to discriminate between the species depends on the **resolution** of the instrument. High resolution instruments usually involve electrostatic and electromagnetic focusing which provide a very accurate analysis of the mass spectrum. However, such instruments are elaborate in design and construction and are not normally used for combined GCMS equipment, largely because scanning speeds are slow. GCMS instruments usually have a low to medium resolution and are based on alternative, more economical methods of ion formation, focusing and collection.

The commonest techniques now used for capillary GCMS involve either **quadrapole** or **ion trap** techniques. These have the necessary advantage of rapid scanning ability and enable the detection of ions having a specific mass to charge ratio (**single ion monitoring**, SIM).

A quadrapole instrument consists of four parallel rods arranged equidistant from each other in a square formation, such that the internal radius equals the smallest radius of curvature of the hypobolic path taken by ions. The rods are operated under radio frequency (RF) and direct current (DC) conditions under which only ions of a single species can traverse the path through the centre of the quadrapole. Other ions are deflected and cannot be detected. Different ions are detected according to the RF and DC conditions, or alternatively a mass spectrum is produced by varying these conditions systematically with time.

A similar function is performed by the more recently developed **ion trap detector**, except that the quadrapole arrangement is replaced by a circular ring electrode situated between ground potential end caps and which is supplied

CAPILLARY INSTRUMENTATION

with an RF voltage. Thermal electrons are generated from a filament and accelerated through a gate which can be sequentially opened and closed electrically. The capillary GC effluent is ionized by the accelerated electrons, in the presence or otherwise of a reagent gas, and the ions formed are held inside the ring electrode (ion trap). The ions are subsequently analysed by increasing the radio frequency of the electrode which enables the ions to exit the trap sequentially and be detected.

Figure 3.23 shows the use of GCMS using selective ion monitoring for the analysis of the tetrachloro-dioxins which is an important environmental application.

In primary investigational studies a full mass spectrum would be run for all the required peaks in the chromatogram to provide a full identification. The combined evidence of the mass spectrum and highly specific retention characteristics of a compound is usually sufficient proof of identity.

3.18.3 Coupling the Column to the Mass Spectrometer

Since the optimum flow rate into the ion source of a mass spectrometer is in the region of $1-2$ ml min^{-1}, direct coupling of a capillary column provides the

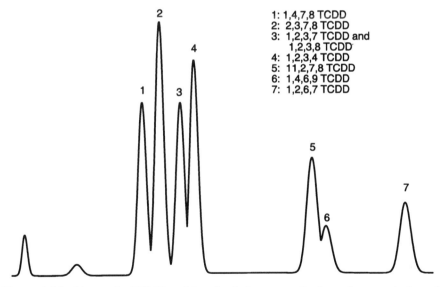

Figure 3.23 Use of GCMS with single-ion monitoring for analysis of tetrachlorodibenzodioxins (TCDDs). Column: 50 m × 0.22 mm (0.2 μm). FSOT column coated with poly(cyanopropylsiloxane); splitless injection; temperature: 45 °C to 190 °C (ballistic), to 240 °C at 5 °C min^{-1}. Mass spectrometer $m/z = 320$ and 322. (Reproduced with permission of James Eichelberger, US Environmental Protection Agency)

simplest form of combination, compared with packed GC or HPLC columns. Thus it is fairly straightforward to feed the capillary column outlet directly into the ion source [27,28]. However, one disadvantage of this method is the high vacuum at the end of the column which can adversely affect column performance, particularly if the column has a low pressure drop as with a wide bore column. In this event a fixed restrictor can be placed between the end of the column and the ion source so that most of the pressure drop below atmospheric pressure takes place in the restrictor. A system of this type will include an adjustable bypass valve to vent large volumes of solvents or corrosive reagents used for derivatization. A third type of coupling for capillary GCMS is referred to as open split coupling [29]. Essentially, this is a simple tube into which the capillary exit and the ion source inlet are fed via smaller capillary tubes which are inserted into each end of the coupling leaving a small gap at each end. The split coupling is continuously purged with a gas stream and the amount of capillary effluent sampled by the mass spectrometer depends on their ratio of column flow rate to ion source entry flow rate.

The combination of capillary GC with mass spectrometry has advanced considerably with the development of dedicated instrumentation and the interested reader is advised to refer to the specialist treatises on the subject [30–33].

3.19 PEAK MEASUREMENT AND DATA PROCESSING

Analytical techniques are discussed in Chapter 8 but these depend on the quantitative interpretation of the detector output which is an analogue signal from the associated amplifier. This signal is sent to further instrumentation for handling in an appropriate manner.

In its simplest form, the instrumentation for peak measurements can consist of a recording device to display the chromatogram as an analogue signal from the detector amplifier. This is then used for all measurements of retention times, peak areas and peak heights. Although such procedures are laborious and limited to the range of the recorder, they can produce acceptable results provided the quality of the chromatography is satisfactory.

Nowadays, these measurements are usually carried out by **data systems** but these should not be regarded as essential to the separation. The quality of the final results still depends on the quality of the column and its operating conditions, and these aspects should never be sacrificed in the misguided belief that expensive data systems can compensate for poor chromatography. Nevertheless, the modern data system is a fact of life, and is indeed essential for the fast and efficient processing of chromatographic results.

Data systems have evolved from the basic need to measure the areas of chromatographic peaks for quantitative analysis. Early integrators were of the mechanical ball and disc type which were simply adjuncts to the recorder and produced a series of oscillations which were proportional to the peak area and had to be counted manually. Another type of integrator involved the use of a potentiometer coupled to the recorder pen movement which produced a voltage proportional to the input voltage to the recorder. This was converted to an digital voltage by an analogue to digital conversion circuit, and values were accumulated for each peak to integrate the area. These methods were still limited by the recorder range however and, consequently, separate integrators which received the detector signal directly from the amplifier without the prior intervention of the recorder were soon developed. However, these early instruments were very dependent on manual settings for baseline position and peak recognition, and bias errors were commonplace. A useful reference source to peak integration methods has been published by Dyson [34].

The advent of the microprocessor revolutionized the GC integrator which rapidly developed into what we know today as the **data system**, although the basic principle of peak integration is still the same, namely, analogue to digital conversion followed by accumulation of the data. Very fast sampling of the peak analogue signal is necessary, however, to achieve the necessary level of accuracy for capillary peaks. We can summarize the main features of a typical data system as follows.

1. Retrieval of stored analytical methods to be used from system memory. This will include calibration data for the components to be analysed.
2. Noise smoothing and/or filtering.
3. Peak detection using derivatives of the peak signal.
4. Storage of the time elapsed from the start to the peak maxima (retention times).
5. Repetitive analogue to digital conversion of the signal voltage for each peak duration and storage of values in system memory. For capillary peaks the sampling frequency must be sufficiently high.
6. Regeneration of the chromatogram and allocation of baseline positions were necessary. This will usually include the retention times.
7. Adjustment of stored digital data to allocated baselines, summation and readout of resulting area counts.
8. Correction of data for stored response factors and calculation of quantitative results in the required format.
9. Printout of analytical data.

This sequence may well vary from one commercial system to another, and there is a variety of possible systems available. The potential user should take considerable care to examine the data systems available and ensure that an appropriate one, which will fulfil the analytical requirements, is chosen. For instance, most data systems can accept raw data from several chromatographs simultaneously, and this would be advantageous for the busy analytical laboratory.

An important aspect of the data system is its ability to handle partial separations. Figure 3.24 shows four possible situations and the type of baseline fitting that should be applied. In the 'ideal' situation, shown in (a), the peaks are Gaussian and completely separated. This should give the maximum accuracy and, obviously, this is the desired objective. However, it is not always possible to achieve such a clear separation between the required components and the appropriate allocation of the peak baseline can be very difficult. For overlapping peaks of similar size, as shown in (b), the **perpendicular drop** method is usually appropriate. Thus the integrator detects the trough minimum between the peaks and treats this as the changeover time. Obviously, this does produce errors of the type discussed in Chapter 2, and whether or not these are acceptable are subject to test. A common situation is shown in (c), particularly in trace analysis where minor components appear on the tail of major components. The **tangential skim** method can be applied automatically by the data system. More recent systems, however, provide an option, as shown in (d). Thus various manual baselines can be tried on the data using computer graphics to find the most acceptable fit.

Another useful feature of many data systems, particularly valuable in qualitative investigations, is the ability to reproduce any portion of the chromatogram with considerable amplification, or to zoom into areas containing minor peaks etc. for further examination. Also stored chromatograms can be retrieved from the memory for comparison purposes by superimposition, etc.

The software for the most recent generation of data systems is increasingly based on the Microsoft Windows® graphical user interface with which most

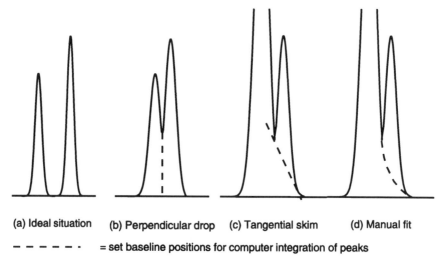

(a) Ideal situation   (b) Perpendicular drop   (c) Tangential skim   (d) Manual fit

– – – – – ·   = set baseline positions for computer integration of peaks

Figure 3.24 Baseline construction methods

Table 3.3 Comparison of detector properties

Detector (type)	Minimum detectability	Linear range	Carrier gases	Application
FID (Mass)	10^{-12}–10^{-13} g C s^{-1}	10^6–10^7	Any	Most organic compounds
NPD (Mass)	N: 2×10^{-13} g N s^{-1} P: 10_{-13} g P s^{-1}	N: 10^5 P: 10^4	Any	Nitrogen and phosphorus compounds
FPD (Mass)	10^{-10}–10^{-11} g S s^{-1} 10^{-12} g P s^{-1}	S: 10^2 (log–log) P: 10^3	Any	Sulphur and phosphorus compounds
ECD (Concn)	10^{-13}–10^{-14} g lindane	10^4	He, Ne$_2$ (Ar/CH$_4$ makeup)	Halogens and specific structural types
TCD (Concn)	10^{-9}–10^{-10} g ml^{-1}	10^4–10^6	H$_2$, He	All compounds (universal)
PID (Concn)	10^{-11} g (benzene)	>10^4	Any	Aromatics polyaromatics phenols, etc.

PC users are familiar. Data output from the chromatography is routed to the PC, whereby the installed software can perform all the required data handling operations. Also with most commercial systems, the operator can input data to allow the setting and adjustment of temperature and flow rates from the keyboard. Alternatively, chromatographic variables may be set via the computer software according to a stored method.

3.20 GENERAL SUMMARY

This chapter has dealt with the essential features of hardware of the capillary GC system. This is a fast developing area, particularly in relation to the involvement of modern data systems. As the memory capabilities of PCs increase, we can expect that retention information will, in the future, be stored for many compounds so that appropriate conditions can be determined rapidly for the analysis.

In general, the modern chromatography is highly reliable in operation and operates at very accurate and precise levels of the variables. Although these are factors that can affect the accuracy of quantitative analysis, it is also essential for the equipment to be handled and operated appropriately to achieve the required quality of performance. A lack of knowledge and poor handling will invariably produce poor results in capillary GC regardless of the quality of the gas chromatography.

REFERENCES

1. Rohwer, E. R, Pretorius. V and Apps P. J. (1986) *HRC and CC*, **9**, 295.
2. Grob, K. (1987) *On-Column Injection in Capillary Gas Chromatography*, Chromatographic Methods Series, Heuthig Verlag, Heidelberg.
3. Dressler, M. (1986) *Selective Gas Chromatographic Detectors*, Elsevier, Amsterdam.
4. Scott, R. P. W. (1957) *Vapour Phase Chromatography*, (ed. D. H. Desty), Butterworths, London, pp. 131–145.
5. Harley, J., Nel, W. and Pretorius, V. (1958) *Nature (London)*, **181**, 117.
6. McWilliam, I. G. and Dewar, R. A. (1958) *Nature (London)*, **181**, 760.
7. McWilliam, I. G. (1983) *Chromatographia*, **17**, 241.
8. Jorgenson, A. D., Picel, K. C. and Stamoudis, V. C. (1990) *Anal. Chem.*, **62**(7) 683.
9. Boćek, P. and Janák, J. (1971) *Chromatogr. Rev.*, **15**(2-3), 111.
10. Novák, J. *et al.* (1970) *J. Chromatogr.*, **51**, 385.
11. Bolton, H. C. and McWilliam, I. G. (1971) *Proc. Roy. Soc., (London)*, **A321**, 361.
12. Ševčik, J., Kaiser, R.E. and Rieder, R. (1976) *J. Chromatogr.*, **126**, 263.
13. Lucero, D. P. (1972) *J.Chromatogr. Sci.*, **10**, 463.
14. Guiffrida, L. (1967) *J. Ass. of. Agric. Chem.*, **47**, 293.
15. Patterson, P. L. (1986) *J. Chromatogr. Sci.*, **24**, 466.
16. Kolb, B., Auer, M. and Poppisil, P. (1977) *J. Chromatogr. Sci.*, **15**, 53.
17. Bombick, D. D. and Allison, J. (1989) *J. Chromatogr. Sci.*, **27**, 612.
18. van de Weijer, P., Zwerver, B. H. and Lynch, R. J. (1988) *Anal. Chem.*, **60**, 1380.
19. Grant, D. W. (1958) *Gas Chromatography 1958*, (ed. D. H. Desty), Butterworths, London, p. 153.
20. Flinn, C. G. and Aue, W. A. (1980) *J. Chromatogr. Sci.*, **18**, 136.
22. Lovelock, J. E. and Lipsky, S. R. (1956) *J. Amer. Chem. Soc.*, **78**, 546.
23. Zlatkis, A. and Poole, C. F. (eds) (1981) *Electron Capture Theory and Practice in Chromatography*, Elsevier, Amsterdam.
24. Chen, E. C. M., Wentworth, W. E., Desai, E. and Batten, C. (1987) *J. Chromatogr.*, **399**, 121.
25. Lovelock, J. E. (1960) *Nature (London)*, **188**, 401.
26. Langhorst, M. L. (1981) *J. Chromatogr. Sci.*, **19**, 98.
27. McFadden, W. H. (1979) *J. Chromatogr. Sci.*, **17**, 2.
28. Rose, K. (1983) *J. Chromatogr.*, **259**, 445.
29. Davies, N. W. (1988) *J. Chromatogr.*, **450**, 388.
30. McFadden. W. H. (1973) *Techniques of Combined Gas Chromatography–Mass Spectrometry: Applications in Organic Analysis*, Wiley, New York.
31. Gudzinowicz, B. J., Gudzinowicz, M. J. and Martin, H. F. Part II (1976) and Part III (1977) *Fundamentals of Integrated Gas Chromatography–Mass Spectrometry*, Dekker, New York.
32. Message, G. M. (1984) *Practical Aspects of Gas Chromatography–Mass Spectrometry*, Wiley, New York.
33. Karasek, F. W. and Clement, R. E. (1988) *Basic Gas Chromatography–Mass Spectrometry, Principles and Techniques*, Elsevier, Amsterdam.
34. Dyson, N. (1990) *Chromatographic Integration Methods*, (ed. R. M. Smith), Royal Society of Chemistry Chromatographic Monographs Series, Royal Society of Chemistry, Cambridge.

4

The Open Tubular Column

4.1 INTRODUCTION

The concept of using a coated open tube for gas chromatography was an exciting departure from the conventional use of packed columns and it inspired a great deal of early interest. Progress was initially very rapid as the inherent advantages of improved resolution and faster separation speed became apparent. There were inherent disadvantages, however, namely, a poor level of reproducibility and a lack of column stability. The stationary phase film thickness was usually unknown and certainly varied along the column. Also the lack of any reliable and consistent source of capillary tubing was a major disadvantage as most commercial sources were unsuited to chromatographic use. Columns constructed from copper, stainless steel, nylon, aluminium and glass gave a degree of success depending on the treatment applied to the material and the type of sample, the nature of the stationary phase and its film thickness. One problem was the presence of 'active', or adsorptive sites on the interior surface of the tubing as this was the primary cause of peak tailing with polar compounds.

For many years high grade stainless steel and glass in its many varieties were the chosen column materials, and a variety of treatments were used to reduce peak tailing effects and to produce acceptable results. The stationary phase was usually applied by the **dynamic coating procedure** which involves passing a solution of the phase in a volatile solvent through the column under an inert gas pressure. Excess solvent is removed from the wet interior surface after the bulk of the solution has emerged by continuing the flow of gas for a period of time. Although this technique works very well, it tends to give a variable and non-reproducible film thickness. This procedure has now largely been replaced by a **static** method of coating (see Section 4.25) although the dynamic method may still be used for some columns in commercial production.

These uncertainties, combined with a lack of reproducibility, delayed any widespread acceptance of the new columns as serious competitors to packed columns. Also, committees that were responsible for standardizing analytical methods showed a clear reluctance to allow their use in standard test procedures.

The situation changed dramatically on the arrival of flexible fused silica as a new column material in 1978 (Chapter 1, reference 42). The majority of commercial open tubular columns are now made from this material, which, with new manufacturing techniques now ensure a high degree of reproducibilty and reliability.

4.2 MANUFACTURE OF FLEXIBLE FUSED SILICA OPEN TUBULAR (FSOT) COLUMNS

Modern FSOT columns are made by a variety of techniques depending on the manufacturer and the type of column. A typical manufacturing sequence would be as follows.

- Capillary drawing to required diameter and coating of the outside surface with a continuous layer of polymer or other suitable material
- Thorough testing of fused silica capillary tube for flexibility and strength
- Cutting to required lengths
- Interior surface roughening (not needed for immobilized phases)
- Surface deactivation
- Coating with stationary phase
- Conditioning
- Performance testing

Another important type of capillary column is the **porous layer open tubular**, or **PLOT** column. This is coated with a layer of particulate adsorbent material and is manufactured in a variety of ways, the details of which are not freely available. PLOT columns are discussed in Chapter 5.

4.2.1 Drawing of FSOT Columns

The early glass columns were drawn down to capillary dimensions using apparatus of the type first described by Desty (Chapter 1, reference 41). A similar principle is now applied to the preparation of fused silica capillary tubing but with rather more sophisticated equipment. The drawing and preparation principle is illustrated by the schematic diagram in Figure 4.1.

A fused silica preform is introduced to the electrical furnace at a constant rate by the feed rollers. This heats the silica to its softening point which in the region of 2000 °C, allowing the drawing rollers to pull the tubing down to the

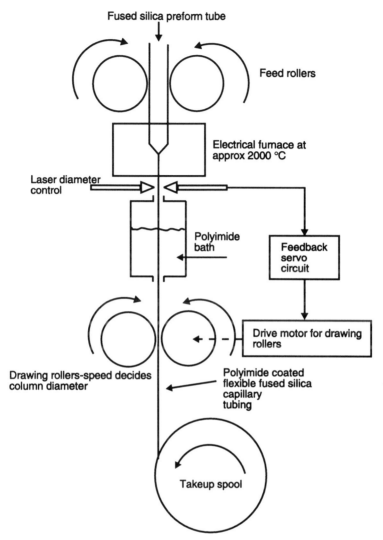

Figure 4.1 Diagram of equipment used for the manufacture of fused silica columns

required diameter. This diameter is maintained constant by the laser activated feedback control circuit. Thus any tendency for the diameter to change causes a compensatory adjustment to the rotational speed of the drawing rollers. Between the furnace and the drawing rollers the capillary passes through several polyimide treatment baths which applies a skin of this polymer to the **exterior** surface.

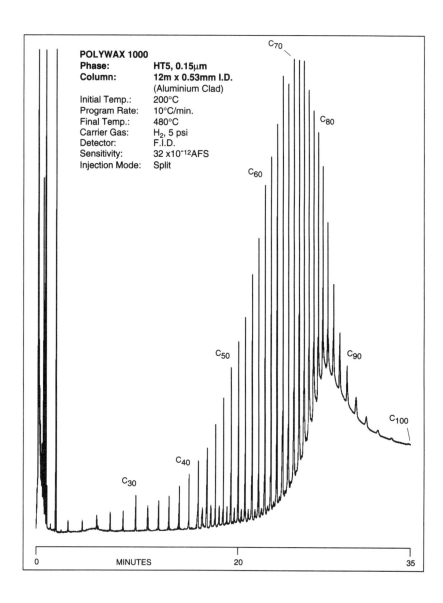

Figure 4.2 Use of aluminium clad fused silica column for high temperature separation of polywax 1000. Column: 12 m × 0.53 mm (aluminium clad) coated with HT5 (0.15 μm); temperature 200 to 480 °C at 10 °C min^{-1}; carrier gas H_2 at 5 psi; split injection; detector FID. (Chromatogram used with permission of SGE International Pty Ltd)

4.2.2 Strength and Durability of FSOT Columns

The exterior polymer coating which is applied during the capillary drawing process is essential to give the final column its flexible characteristics, and **polyimide** has been found to be very suitable for this purpose. The quality of this outer coating is critical as any flaws or fissures will cause the tube to fracture very easily. It is important for users to be aware of the vulnerability of this exterior coating and not to subject it to any treatment that could cause damage. The polyimide layer is stable up to about 400 °C after which it will start to degrade.

In recent years, stationary phases have been developed which are capable of operating at temperatures higher than 400 °C, i.e. higher than the operating range of polyimide coated FSOT columns. Some manufacturers advocate the use of **aluminium clad** fused silica columns for high temperature applications, an example of which is shown in Figure 4.2 for a sample of Polywax 1000™. This is a polyethylene, covering a carbon number range of more than C100. The stationary phase used for this separation is a carborane modified polysiloxane bonded phase which is stable up 480 °C.

An alternative to the use of aluminium coatings on fused silica is to revert to the use of stainless steel. Modern stainless steel columns are manufactured using patented techniques which avoid many of the earlier problems (see Section 4.4).

4.2.3 Effects of Inner Surface

Open tubular columns were originally coated with the stationary phase without any pretreatment of the interior surface. While this was sometimes satisfactory, many phases formed minute droplets, particularly if thick films were attempted. This effect is shown diagrammatically in Figure 4.3a.

The propensity of the stationary phase to form droplets rather than a uniform cohesive layer depends on the balance between the interfacial surface tension of the phase and the tube surface, and the intrinsic surface tension of the phase itself. Polar liquids, which usually have higher surface tensions than non-polar ones, are therefore more likely to give this problem when they come into contact with the smooth surface of the tube. Early columns were consequently more successfully coated with non-polar liquids. Droplet formation results in a poor column performance because of slower mass transfer and adsorption at uncovered active sites between the droplets.

The other major problem mentioned earlier, namely, the presence of active sites on the inner surface, is illustrated by Figure 4.3b. Active sites usually consist of silanol groups at the fused silica surface which can interact with the polar groups in the sample molecules thereby creating an additional retentive force. Non-linear adsorption at these sites causes the peaks to tail and this will reduce the resolution, chromatographic sensitivity and analytical accuracy.

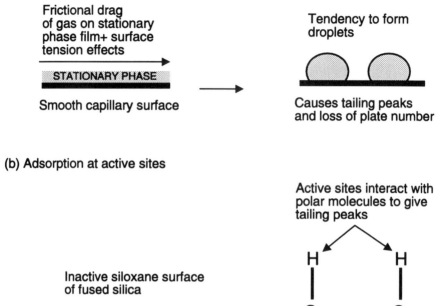

Figure 4.3 Reasons for surface pretreatment

Active sites can also arise from the presence of metal oxides in metal columns, and metallic ions and hydroxyl groups in glass columns.

Non-polar compounds should give symmetrical peaks because peak tailing is associated with polar interactions with active sites. Aliphatic hydrocarbons, for instance, are unlikely to give peak tailing whatever type of column is installed. If peak tailing is experienced with such compounds then the user should suspect other causes such as poor column installation, poor sample injection, or extra-column volume effects as described in Chapter 3. Compounds that may give tailing peaks are amines, carboxylic acids, amides, phenols and diols, particularly if there is more than one such functional group in the molecule. In extreme cases, the compounds may even be irreversibly retained by the column. Medium polarity compounds such as ketones, aldehydes and alcohols may also give some tailing on undeactivated columns.

THE OPEN TUBULAR COLUMN

Figure 4.4 illustrates typical peak tailing arising from the presence of active sites.

Modern fused silica consists of pure SiO_2 and is manufactured by the oxidation of pure silicon tetrachloride; consequently, there is no significant contamination by metallic oxides, etc. It is therefore comparatively inert, with the advantage that it has a known and exact composition. The surface consists largely of siloxane groups as shown in Figure 4.3b, but as we have noted, there is a degree of activity due to the presence of silanol groups. An effective deactivation treatment is therefore necessary during manufacture.

4.2.4 Surface Pretreatment of Open Tubular Columns

The effective elimination of the problems of poor wettability and non-linear adsorption effects in open tubular columns has preoccupied research efforts for several decades. Techniques were often developed by chromatographers who made columns in their own laboratories; a common activity during the early development of the technique. It is rarely practised today because of the high quality and availability of commercial columns. The methods employed by modern manufacturers are reliable and highly effective, but so far no method of treatment can be claimed to be one hundred per cent successful. Also, many of the aspects of column manufacture differ considerably from one manufacturer to another and involve unpublicized processes. From the user's viewpoint,

(a) Polar compounds on deactivated column (simulated chromatogram)

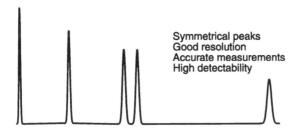

Symmetrical peaks
Good resolution
Accurate measurements
High detectability

(b) Polar compounds on non-deactivated column (simulated)

Tailing peaks
Poor resolution
Inaccurate measurements
Poor detectability
Retention times may be affected

Figure 4.4 Effect of active sites

it is important to be able to choose the most appropriate column for the required application from the range available, particularly if very polar compounds are involved.

When most open tubular columns were constructed from various forms of glass a variety of surface roughening treatments were applied to improve wettability. These treatments were usually based on surface etching with HCl in liquid or gaseous form, [1,2] usually under pressure at elevated temperatures. Alternatively, colloidal particles were deposited as a way of extending the surface. The latter is one of the current methods employed now for FSOT columns, and can involve either sodium chloride or barium carbonate microcrystals. Sodium chloride crystals are deposited over the inner surface of the capillary tube from a non-aqueous colloidal suspension [3]. The alternative use of a layer of barium carbonate crystals was first described by Grob [2]. The layer is formed by chemical reaction *in situ* with CO_2 gas after coating the interior surface of the column dynamically with a solution of barium hydroxide. Stable layers of various thicknesses can be formed in this way. A dense layer is suitable for coating polar phases and a thinner coating is preferred for non-polar phases.

The second major problem mentioned above, i.e., that of surface activity from active sites is perhaps a more difficult one to overcome because all column materials have some degree of residual activity which can cause tailing peaks whatever deactivation treatment is applied. The practical objective is to apply the most effective deactivation treatment possible to enable a wide variety of compounds to be separated without peak tailing. Considerable problems were encountered with early glass columns because of the variable nature of glass, its non-stoichiometric composition and the probability of contamination by metallic ions. Fused silica has a defined composition, but even this material can have variable surface properties depending on the density and configuration of the residual silanol groups. Nevertheless, it is far more inert than most other materials and it can be deactivated very effectively.

Deactivation treatments are based on the interaction of reagent molecules with active sites to effectively 'neutralize' them, or the application of a skin over the surface to cover the active sites. One treatment that was used for glass columns involved silylation to convert surface silanol groups to their trimethylsilyl ether derivatives using reagents such as trimethylsilyl chloride or hexamethyldisilazane. These reactions can be simply represented as:

$$—Si—OH + (CH_3)_3—Si—Cl = Si—O—Si—(CH_3)_3 + HCl \quad (4.1a)$$
surface silanol group + reagent = trimethylsilyl group

A common method of treatment for FSOT columns is high temperature gas phase deactivation with polysiloxanes using a procedure described by Schomburg and coworkers [4]. The precise mechanism of this treatment is

THE OPEN TUBULAR COLUMN

uncertain but probably involves the initial decomposition of the polysiloxane into highly reactive products which then interact with the surface silanol groups to give a surface coverage of polysiloxane polymer. Columns deactivated in this way are normally used with low polarity phases and for high temperature operation.

Another effective treatment for FSOT columns involves gas phase deactivation with a high molecular weight polyethylene glycol (PEG). Again the probable mechanism involves the bonding of surface silanol groups with the PEG degradation products. PEG deactivation can be carried out in a variety of ways but it is very effective particularly for the subsequent coating of polar stationary phases [5].

In spite of the efforts of manufacturers to produce columns that give minimal tailing for polar compounds, successful results will still not be obtained unless the user chooses the stationary phase and film thickness with care and foresight. Fortunately, many difficult applications can be performed with specially designed columns that are guaranteed to be successful by the manufacturer. Typical applications of this type include many types of environmental analyses such as pesticides, halocarbons, PCBs, dioxins, amines, triglycerides, gasoline, polycyclic hydrocarbons, optical isomers, glycols, alcohols, fatty acid methyl esters (FAME) and many others. One example is given in Figure 4.5 showing the separation of free fatty acids on a PEG deactivated column coated with FFAP (free fatty acid phase).

When a decision has to be made by the user for an appropriate stationary phase the usual advice is to choose a phase which has similar polarity characteristics to the sample (see Section 4.8). Polar stationary phases will tend to 'mask' the effects of active sites thereby reducing their adsorptive effects. Another way of reducing tailing is to choose a thicker stationary phase film as described in Section 4.7.3 as this also has the effect of covering the active sites.

4.2.5 Application of the Stationary Phase

The dynamic coating procedure described earlier was not very satisfactory because of its poor reproducibility and uncertainty with regard to the final film thickness. A static method is now usually employed which gives a uniform deposit of stationary phase with a known film thickness.

For the static coating method, a known concentration of the stationary phase in a volatile solvent is used to completely fill a preweighed column. After reweighing, one end of the column is sealed and a vacuum is applied to the open end to remove the solvent by its vapour pressure which leaves the residual stationary phase deposited evenly on the column wall. The column diameter is calculated from the weight and density of the solution used to fill it, and the film thickness is computed from the mass and density of stationary phase.

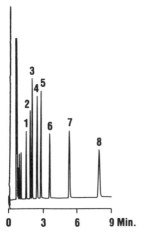

1. Acetic acid
2. Propionic acid
3. Isobutyric acid
4. n-Butyric acid
5. Isovaleric acid
6. n-Valeric acid
7. Caproic acid
8. Heptanoic acid

Figure 4.5 Chromatogram of free short chain fatty acids. Column: 15 m × 0.53 mm Econo-Cap™ coated with FFAP (1.2 μ). Temperature: 50 °C; carrier gas helium; detector FID. (Reproduced by permission of Alltech Associates Applied Science Ltd)

4.2.6 Column Conditioning

It is essential for the final column to possess the requisite selectivity for the chosen stationary phase and to operate reliably and reproducibly up to a specified temperature limit. This specified limit is known as the **maximum allowable operating temperature**, or MAOT value, and depends mainly on the type of phase and the mode of operation. The MAOT value is the maximum column operating temperature for extended use of the column as recommended by the manufacturer. All columns have a finite life span for a variety of reasons and they do begin to deteriorate from the moment they are put into service. However, a minimum life span of at least six months should be expected. Much depends on user care and the conditions, especially the temperature, to which the column is subjected. Manufacturers, usually quote the MAOT value in a meaningful way, and one method is to define it in terms of the half-life of the column expressed as the maximum temperature under optimum flow conditions at which the column will retain one half of its stationary phase for a period of six months. Two MAOT values are usually specified, relating to the use of isothermal and programmed temperature operation respectively. Values for programmed operation are slightly higher than for isothermal operation because the column is normally cooled shortly after reaching the maximum temperature of the programme. Typical values for some common stationary phases are given in Table 4.9.

In order to achieve stable MAOT values for the final column, it is essential to remove all volatile impurities, residual solvent, etc. This is achieved by the process of **conditioning** which is the final stage of manufacture apart from chromatographic testing. Column conditioning involves passing a slow stream of inert gas through the column at a temperature which is successively increased to the MAOT. The gas flow is then maintained for a number of hours, after which the column is tested for its chromatographic properties, particularly retention factors, theoretical plate number and coating efficiency.

Although all commercial columns are conditioned by the manufacturer the purchaser will find that it is invariably necessary to carry out a further period of conditioning before stable operation is achieved. This also applies to existing columns that have been reinstalled after a period of storage.

4.3 IMMOBILIZED AND BONDED STATIONARY PHASES

Even after the various surface roughening procedures have been applied, the stationary phase stability still depends on surface tension forces and so it can easily be disturbed either by maltreatment or by solvents from the sample, etc. Not surprisingly, therefore, a considerable research effort has been applied to finding ways of stabilizing the stationary phase. The IUPAC definition of an immobilized phases is one that has been stabilized by *in situ* polymerization either during or immediately following the coating procedure. Conversely, **bonded phases** are defined as stationary phases which are stabilized by covalent linking to the interior wall of the column. Manufacturers rarely adhere to these definitions and descriptions such as 'bonded' or 'chemically bonded' are commonplace. For practical convenience, both types of column will be referred to as 'CB' columns, and the phases as 'immobilized'.

Let us define an immobilized phase, however it is produced, and whatever is the mechanism of stabilization, as one which cannot be removed by common solvents and yet gives similar retention and selectivity properties as the original phase. Immobilization does **not** imply that the phase has a much higher MAOT, although there may be some marginal improvements. As with other aspects of column manufacture there is a great deal of secrecy in relation to the specific methods used for immobilization, but they are limited at the present time to various polymers, particularly those based on polysiloxanes where immobilization depends on further polymerization to higher molecular weight products.

Immobilization is normally carried out during the column coating procedure either by thermal treatment or by the use of free radicals derived from a suitable reagent such as a peroxide, or by irradiation. One method is to first deactivate the column using Schomberg's polysiloxane deactivation treatment described in Section 4.2.4, and then to coat the column using the static method. No surface roughening is necessary for immobilized phases. The 'chemical bonding' procedure is then applied following the removal of the solvent, and usually involves sealing the two ends of the column and raising its temperature to 'cure' the phase. Impurities, excess reagent, etc. are then removed from the column by the passage of various solvents.

An example of the use of a chemically bonded column is given in Figure 4.6 which shows the separation of antidepressants on a bonded 7% methyl 7% cyanopropyl siloxane phase.

The two major advantages of immobilization phases are that they are no longer soluble in normal solvents, and much thicker layers can be applied to the column. The first of these advantages enables columns to be regenerated by solvent rinsing to remove soluble contaminants which accumulate in the column and cause a loss of performance. This regeneration procedure involves removing the column from the GC equipment and flushing it with an appropriate volume of a solvent such as **methylene chloride, tetrahydrofuran**, or even **water** if appropriate. This is carried out with a syringe filled with the chosen solvent and connected to one end of the column. The excess solvent is removed and the column dried in a slow current of pure, dry, inert gas.

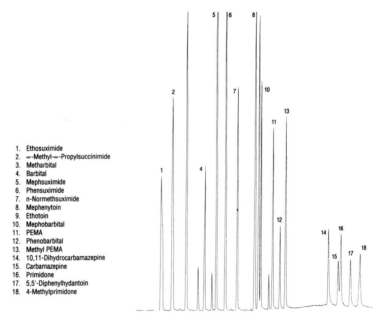

1. Ethosuximide
2. ∝-Methyl-∝-Propylsuccinimide
3. Metharbital
4. Barbital
5. Mephsuximide
6. Phensuximide
7. n-Normethsuximide
8. Mephenytoin
9. Ethotoin
10. Mephobarbital
11. PEMA
12. Phenobarbital
13. Methyl PEMA
14. 10,11-Dihydrocarbamazepine
15. Carbamazepine
16. Primidone
17. 5,5'-Diphenylhydantoin
18. 4-Methylprimidone

Figure 4.6 Separation of pharmaceuticals on a CB column. Column: 30 m × 0.25 mm bonded methyl 7% cyanopropyl 7% phenyl silicone, film thickness: 0.25 μm; temperature: 145 to 290 at 12 °C min^{-1}. Detector: mass selective detector (MSD). (Reproduced by permission of Quadrex Corporation)

THE OPEN TUBULAR COLUMN

Although this rinsing procedure is usually successful in prolonging the effective lifetime of columns, they invariably deteriorate gradually with use and must be replaced in due course.

Another important feature of CB columns, arising from the insolubility of the phase in solvents, is their use in connection with sample introduction (Chapter 6). Splitless and on-column injection involve the presence of liquid sample inside the column which would dissolve any non-bonded phase and destroy the column.

Phase immobilization is very important to the modern technique in enabling thick layers of stationary phase to be applied to the column. Without phase immobilization the maximum stable film thickness is about 0.4 μm, thus imposing a severe limitation with regard to the separation of very volatile compounds. Immobilized phases, however, can be coated with film thicknesses of up to several micrometres which are sufficient to enable the separation of compounds such as freons, hydrocarbon gases, solvents, etc.

4.4 STAINLESS STEEL OPEN TUBULAR COLUMNS

As was mentioned earlier, stainless steel was one of the original materials used for open tubular columns. However, it was not very satisfactory for polar compounds, tending to produce tailing peaks in many cases. Considerable progress has been made in recent years in the manufacture of stainless steel columns particularly for high temperature use. Descriptions of the manufacturing techniques are not available for obvious reasons but they are all based on applying several impermeable and inactive layers to the interior surface of the tube. The development of these columns has proceeded hand in hand with that of more thermally stable stationary phases based on polysiloxane polymers.

Examples of the use of these columns are given in Figure 4.7.

4.5 PERMEABILITY OF OPEN TUBULAR COLUMNS

According to equation (2.8) the column exit gas velocity is given by,

$$u_0 = \frac{B_0(p_i^2 - p_i^2)}{2\eta L p_0} \tag{4.1}$$

but $\bar{u} = u_0 j$, where j is the compressibility correction factor (see Chapter 2), i.e.

$$\bar{u} = \frac{3(P^2 - 1)B_0(p_i^2 - p_0^2)}{4(P^3 - 1)\eta L p_0} \tag{4.2}$$

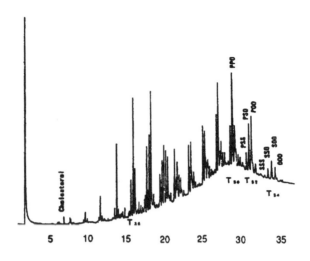

Peak identification:
1. naphthalene
2. acenaphthylene
3. acenaphthene
4. fluorene
5. phenanthrene
6. anthracene
7. fluoranthene
8. pyrene
9. benz[a]anthracene
10. chrysene
11. benzo[b]fluoranthene
12. benzo[k]fluoranthene
13. benzo[a]pyrene
14. indeno[1,2,3-cd]pyrene
15. dibenz[a,h]anthracene
16. benzo[g,h,i]perylene

Figure 4.7 Examples of separations using stainless steel columns. (a) Separation of butter triglycerides. Column: 30 m × 0.25 mm Ultra ALLOY-TRG™ coated with 0.15 μm film 'triglyceride' phase. Temperature: 250 to 360 °C at 4 °C min^{-1}. (Reproduced by permission of Quadrex Corporation.) (b) Separation of polycyclic aromatic hydrocarbons (PAH) according to EPA610. Column: 25 m × 0.25 mm Ultimetal™ coated with polysiloxane (CPsil PAH CB 0.12 μm). Temperature: 70 to 300 °C at 3 °C min^{-1}; carrier gas H$_2$ 30 cm s^{-1}; split injection; detector FID. (Reproduced by permission of Chrompack International BV)

THE OPEN TUBULAR COLUMN

This simplifies to,

$$\bar{u} = \frac{3B_0 \Delta p (P+1)^2}{4\eta L (P^2 + P + 1)} \qquad (4.3)$$

where $\Delta p = p_i - p_o$ (pressure drop) and $P = p_i/p_o$.

The function $(P+1)^2/(P^2+P+1)$ decreases slowly over the range of pressures applicable to open tubular columns, varying from 1.33 for $P = 1$ ($\Delta p = 0$), to 1.29 for $P = 2$ ($\Delta p = 1$). Assuming a mean value of 1.3 for this function,

$$\bar{u} \cong \frac{B_0 \Delta p}{\eta L} \qquad (4.4)$$

where Δp is expressed in dyn cm^{-2}, L in cm, η in poise and B_0 in cm^2 (1 dyn = 10 μN; 1 poise = 0.1 Pa s)

The specific permeability for an open tubular column is accurately calculated from its diameter from the Poiseuille equation, namely,

$$B_0 = \frac{d_i^2}{32} \qquad (4.5)$$

substituting this value of B_0 into (4.4)

$$\bar{u} = \frac{d_i^2 \Delta p}{32 \eta L} \qquad (4.6)$$

A comparison can be made with packed columns where the relationship between the pressure drop and gas velocity is given by the Kozeny–Carman equation [6]. This is similar to (4.6) except for the inclusion of the porosity factor ε in the denominator. Also for packed columns, $B_2 = d^2{}_p/1000$, d_p being the mean particle diameter of the packing material [7].

Table 4.1 gives calculated values of specific permeabilites for packed and capillary columns, and the maximum length of column that could be used at optimum gas velocity assuming a maximum pressure drop of 5 bar. The table includes values for **packed capillary columns** for comparison. This type of column is usually loosely packed with diatomaceous support material and then drawn down to capillary dimensions. The loose packing results in permeabilities which are intermediate between those of true packed and open tubular columns although this technique does not preclude the possibility of using other means of packing small diameter columns.

Clearly it is not possible to employ very long packed columns in conventional GC equipment because of the extremely high pressures that would be required. Even relatively short columns require a higher pressure drop than for the average open tubular column and so there is a larger velocity gradient along the column

Table 4.1 Comparison of column permeabilities and maximum column lengths

Column	Maximum length (m) to produce average linear velocity of He	Specific permeability $B_0(\times 10^6)$cm^2	Approximate maximum theoretical plate number
Packed with 60/80 mesh diatomaceous support	9 (25 cm s^{-1})	0.44	10 000–15 000
Packed capillary			100 000–150 000
0.55 mm 60/70 mesh*	100 (25 cm s^{-1})	22.38	
0.38 mm 85/90 mesh†	100 (ditto)	2.1	
Open tubular 0.1 mm i.d.	10 (60 cm s^{-1})	3.1	100 000
Open tubular 0.15 mm i.d.	30 (40 cm s^{-1})	7	200 000
Open tubular 0.25 mm i.d.	160 (25 cm s^{-1})	19.5	640 000
Open tubular 0.32 mm i.d.	300 (20 cm s^{-1})	32	1 000 000
Open tubular 0.53 mm i.d.	1500 (12 cm s^{-1})	88	2 800 000

*Landault, C. and Guiochon, G. (1965) *Gas Chromatography 1964*, (ed. A. Goldup), The Institute of Petroleum, London, pp. 121–139.
† Grant, D. W. (1966) *Proc. 4th Wilkins Gas Chromatography Symposium*, Varian Aerograph AG, University of Manchester.

length. According to the van Deemter equation for packed columns (2.46), and the Golay theory (2.56) the **optimum** gas velocity is directly proportional to the gaseous interdiffusion coefficient D_g when the gas phase mass transfer coefficient C_g is predominant. D_g, however, is inversely proportional to the gas pressure and so the optimum velocity also changes with distance along the column, and this would tend to compensate for the effect of the velocity gradient. The Giddings' modification of the Golay equation (2.51) allows for this effect by its use of the gas diffusion coefficient at the column exit pressure, and by incorporating functions of the pressure drop. A precise evaluation of the overall effect is complex but if $C_g \gg C_s$, and the column is operating at its optimum **average** carrier gas velocity then all parts of the column should also be close to optimum, regardless of the pressure drop. This would be the situation with regard to packed columns containing very low loadings of the stationary phase, and with thin film capillary columns. Most packed columns have lower phase ratios than capillary columns and so the liquid phase mass transfer coefficient is usually the predominant factor. Thus considerable regions of the column are likely to deviate from the optimum velocity at high pressure drops. The major consequences are that theoretical plate numbers are lower than expectation and that their relationship with column length is non-linear.

An interesting theoretical aspect applies to the use of ultrahigh pressure drops. Figure 2.1 indicates that the gas velocity flow profile becomes almost constant except for a very rapid increase over the final fraction of the column. Provided that the **inlet** gas velocity is at optimum then most of the column would also be close to the optimum. Because of the very high pressures required to achieve this situation, however, specialized equipment and handling technology would be necessary. Giddings [8] has in fact suggested that the use of ultrahigh pressures in GC could produce some interesting benefits and there is no technical reason why suitable equipment should not be designed. These possibilities do not seem to have been pursued, probably because of the rapid development of capillary GC and the currently accepted status of open tubular columns.

The data shows that much longer open tubular columns can be operated for the same order of pressure drop as used for packed columns unless the diameter is very small. Thus for column diameters of 0.1 mm the pressure drop becomes high and the compressibility effect becomes similar to that with packed columns. Note that the maximum achievable plate number increases rapidly as the column diameter **increases**. This is because the permeability increases with the **square** of the diameter while the plate height decreases only in **direct** proportion to the diameter. Consequently, the pressure drop that would be required to produce a constant plate number decreases with increasing diameter in spite of the proportionate increase in the necessary column length. Of course, this increase in column length means that a considerable price is paid in terms of the longer analysis time.

Table 4.2 gives calculated values of pressure drops for the three most commonly used carrier gases necessary to achieve working average gas velocities in 25 m open tubular columns. The pressure required for other lengths, provided that they are less than the maximum values given in Table 4.1 can be derived by simple proportion. Note the very low pressures required for the wide diameter columns, which are often too low to register on the

Table 4.2 Pressures required for open tubular columns

Column dimensions	Nitrogen $\eta = 2.1 \times 10^{-4}$ poise		Helium $\eta = 2.3 \times 10^{-4}$ poise		Hydrogen $\eta = 1.0 \times 10^{-4}$ poise	
	ΔP(bar)	u(cm s^{-1})	ΔP(bar)	u(cm s^{-1})	ΔP(bar)	u(cm s^{-1})
25 m × 0.25 mm	0.3–0.6	9–18	0.8–1.5	25–50	0.5–1.0	35–70
25 m × 0.32 mm	0.15–0.3	7–14	0.4–0.8	20–40	0.25–0.5	30–60
25 m × 0.53 mm						
(optimum)	0.02–0.04	4–8	0.05–0.1	10–15	0.03–0.06	15–30
(fast mode)	>0.1	>30	0.3–1.0	50–>100	0.2–1.0	>100

Fast mode = the most usual mode of operation for this type of column, producing exit flow rates of 4 ml/min^{-1} for nitrogen to >10 ml/min^{-1} for hydrogen. 1 bar = 1 × 10^5 Newtons m^{-2} = 100 KPa (kilo pascals).

pressure gauge. Modern equipment should be provided with ways of dealing with these very low pressures, possibly by including an optional flow restrictor which is situated at the column exit.

4.6 PARAMETERS OF WCOT COLUMNS AND THEIR EFFECTS

The simple geometry of the open tubular column limits the number of its quantitative parameters to **length, internal diameter**, and **stationary phase film thickness**. The choice of stationary phase is the only qualitative parameter and, consequently, is the least predictable with regard to its effects on a proposed separation. The **sample capacity of the column** is related to its geometry and is considered first.

4.6.1 Sample Capacity of WCOT Columns

A pragmatic view of the capacity of any column is simply the maximum amount of sample that can be applied before the analysis becomes untenable. This very subjective concept would depend on the nature of the sample, the column, its operating conditions, and the sample injection method. Defining capacity in this way may even allow the acceptance of very distorted peaks if the analytical criteria specified by the method are satisfied. However, for the purposes of comparison, and to facilitate the calculation of appropriate sample dilution and injection conditions, a more specific definition of column capacity is desirable. We can then ensure that sufficient, but not excess, quantities of the components enter the column. An acceptable definition is based on loss of plate number resulting from column overload, namely, the maximum mass of a named compound that can be analysed by chromatography under specified conditions before the theoretical plate number falls by a specific amount. There is some disagreement with regard to what constitutes an appropriate amount in this definition. The resolution equation shows that a fall of 10% in the theoretical plate number will only reduce the resolution by about 5% which would be hardly noticeable in most situations. A more reasonable value is a 20% loss of plate number, equivalent to a 10% decline in resolution.

If the mass of any solute exceeds the column capacity then column overload occurs because of deviations from the linear relationship between the solute concentrations in the two phases at equilibrium. According to **Henry's law** (see Section 4.8.1) this linear relationship is normal for dilute solutions and is the basis of the Gaussian peak shape. WCOT columns have high phase ratios and so deviations from Henry's law will normally be due to overloading of the stationary phase as this is the minority phase. This causes an excess of solute to be present in the carrier gas resulting in a lowering of the retention factor. Thus the higher concentration regions will move ahead of the lower concentration regions in the early part of the column causing a distortion of the zone, as illustrated in Figure 4.8a. This type of peak asymmetry is called 'peak fronting' and is an easily recognized effect.

(a) Gas–liquid chromatography (WCOT)

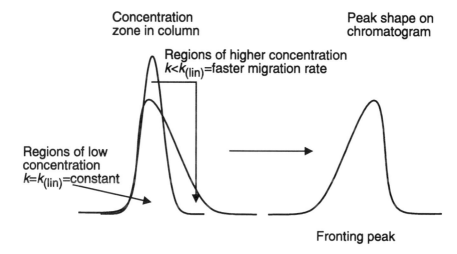

(b) Gas–solid chromatography (PLOT)

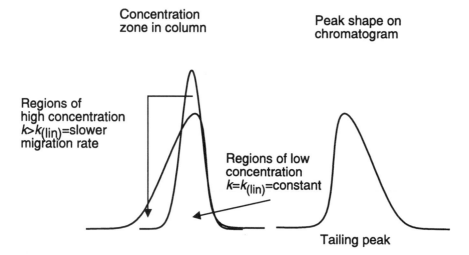

Figure 4.8 Effect of mass overloading on peak shape

Overloading of PLOT columns produces the opposite effect and gives tailing peaks. The reason for this is that the absorbent layer has a much higher capacity for sample than the thin liquid film in WCOT columns. The carrier gas, therefore, becomes overloaded before the stationary phase thereby causing the excess concentration of material to be deposited in the stationary phase and so lag behind the main sample zone. This effect is shown in Figure 4.8b.

Any theoretical treatment of column capacity tends to be unrewarding to the analyst who generally prefers to use working guidelines. The situation is complicated by some ambiguity in the way that capacity has been dealt with in the past. As originally proposed by Keulemans [9], column capacity is the maximum volume of sample **vapour** that can be applied to the column, based on the length of time it takes for the sample to enter the column. The individual **masses** of the sample components are irrelevant to this volume effect provided they do not cause deviations from Henry's law. This aspect of capacity is, of course, extremely important in capillary GC in relation to the method of sample introduction, as discussed in Chapter 6. To avoid any confusion this 'feed time' aspect will be referred to as the **volume capacity** of the column. In discussing the **mass capacity** it will be assumed that the mass of each component is applied to the column quickly and well within the volume capacity.

A practical theory of mass capacity which is generally supported by experimental results, and hence can be used to produce working guidelines, is as follows.

Let a mass W of a particular compound be applied to the first theoretical plate in the column, then

$$W = W_G + W_S \tag{4.7}$$

where W_G = mass of compound in the gas phase and W_s = mass of compound in the stationary phase at equilibrium.

$$\text{but } k = \frac{W_S}{W_G} \tag{4.8}$$

therefore

$$W = W_S\left(1 + \frac{1}{k}\right) \tag{4.9}$$

Let us assume that overloading takes place when the concentration of solute in the stationary phase exceeds a certain critical value S_c. This must, however, be related to the column contribution to zone dispersion which is measured by the standard deviation of the zone width at the end of the column, i.e. $\sqrt{N}$. In other words, the use of very long columns having high plate numbers will tend to compensate for specific amounts of overloading at the start of the column.

THE OPEN TUBULAR COLUMN

Let us therefore define the criterion for significant overloading as a factor $a = S_c/\sqrt{N}$. The concentration of the solute in the stationary phase contained in the first theoretical plate is

$$S_C = \frac{W_s}{v_s} \tag{4.10}$$

where v_s = volume of stationary phase in one theoretical plate, i.e.

$$a = \frac{W_s}{v_s\sqrt{N}} \tag{4.11}$$

$$= \frac{W}{v_s(1 + 1/k)\sqrt{N}} \tag{4.12}$$

$$W_{max} = av_s(1 + 1/k)\sqrt{N} \tag{4.13}$$

where W_{max} = column capacity for the component of interest. But $v_s = H \times$ cross-sectional area occupied by the stationary phase, i.e.

$$v_s = \frac{H\pi d_i d_f}{2} \tag{4.14}$$

$$= \frac{L\pi d_i d_f}{2N} \tag{4.15}$$

therefore, substituting (4.15) into (4.13)

$$W_{max} = \frac{\pi a d_i d_f L(1 + 1/k)}{2\sqrt{N}} \tag{4.16}$$

The sample capacity is therefore proportional to both the film thickness and column diameter. Also, since N is proportional to L the nett effect of increasing the column length is to increase the column capacity in proportion to $\sqrt{L}$. A further conclusion is that the capacity decreases slowly with increasing retention factor k. These conclusions are well supported by practical experience. Figure 4.9 shows some experimental results obtained for a homologous series of n-alkanes.

This decrease in capacity with increasing molecular weight agrees well with the predicted theoretical decrease calculated by multiplying the experimental capacity for n-decane by the function $(1 + 1/k)$, as shown by the points included in Figure 4.9. This is because higher molecular weight compounds occupy a larger fraction of the stationary phase as retention increases (because of the higher partition coefficients) and so they will overload it more readily.

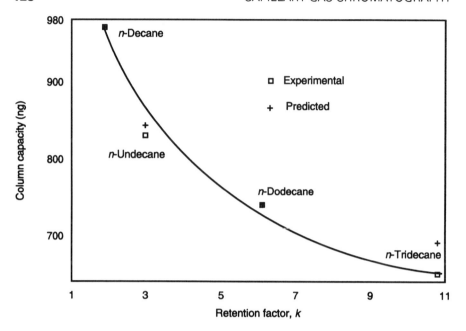

Figure 4.9 Effect of retention factor on column capacity. Column: 25 m × 0.53 mm (2 μm) polymethylsiloxane; temperature 100 °C

This effect is often seen in practice, i.e. overloading becomes more apparent with increasing retention. The value of a in (4.16) was calculated from the experimental data and found to be in the region of 9×10^{-4}. The data given in Table 4.3 using this value for a gives some indication of the mass capacities of open tubular columns.

It is emphasized that the data in Table 4.3 is given for guidance purposes only because variations will occur in practice depending on the nature of both the stationary phase and the sample. A wide discrepancy in polarity for example between the stationary phase and sample components will reduce the

Table 4.3 Nominal sample capacities (ng) of wall coated open tubular columns

$d_f(\mu m)$	Column diameter (mm)				
	0.1	0.15	0.25	0.32	0.53
0.1	1–5	5–10	10–20	15–25	20–100
0.5	5–25	20–50	50–100	80–150	100–300
1	–	–	100–200	150–250	200–600
2	–	–	200–400	300–400	500–1000
5	–	–	500–1000	800–1500	1000–3000

capacity considerably. Such effects are often seen, for instance, when a nonpolar phase such as poly(methylsiloxane) is used for the separation of highly polar compounds such as carboxylic acids.

4.7 QUANTITATIVE PARAMETERS AND RESOLUTION

The theoretical relationships between column geometry and performance as described in Chapter 2, are well supported by practical experience. Let us now see how this theory can be put into practice for everyday use. Equation (2.69), commonly known as the **resolution equation**, is an approximation that is applicable to closely eluting peaks, and is one of the most important practical equations in column chromatography because it provides a working basis for column choice and the setting of operating conditions. Table 4.4 sets out the effects of the three factors in this equation.

4.7.1 Choice of Column Length

As discussed earlier an open tubular column operates at low pressure drops and so we can expect its theoretical plate number to be directly proportional to its length provided that its diameter and film thickness remain constant. The resolution between any two components is proportional to the square root of the plate number and so increasing the column length has only a weak effect. For instance, doubling the column length increases the resolution by a factor of $\sqrt{2}$, which means that (a) there is little or no point in increasing column length by small fractional amounts as it would have an almost negligible effect

Table 4.4 Column resolution effects

Function	Chromatographic variables affecting function	Magnitude of variable effect on resolution
$\sqrt{N}$	Column length	Weak
	Column diameter	Weak
	Film thickness	Weak
	Column temperature	Weak
$k/(1+k)$	Film thickness	Strong for low k values
		Weak for high k values
	Temperature	Very strong for low k values
	Column diameter	As for film thickness
$\alpha - 1$	Stationary phase	Always very strong
	Column temperature	Usually weak, sometimes unpredictable

on the resolution, and (b) there is equally no point in doubling the column length if the resolution between the most critical pairs of compounds is less than unity. Ignorance of this effect can result in a considerable waste of time and expense, as illustrated in Figure 4.10. This shows the result of replacing a 25 m by a 50 m column in an ill-advised attempt to achieve baseline resolution between the two compounds shown. The compounds would still not be resolved to baseline, the separation time would be doubled, and the peak heights would be smaller.

Some guidance for the initial choice of column length is given in Table 4.5 and there should be little need to depart from these recommendations. Thus, if the resolution is not sufficient on the chosen column it is far more effective to use one or more of the other means described in this chapter to improve it.

The recommendations given in Table 4.5 have been calculated according to the complexity of the sample. A useful concept in this context is the **peak capacity** (C_p) of the column, defined as the maximum number of baseline separated peaks that can be accommodated theoretically within a defined elution range. The general calculation for peak capacity is:

$$C_p = \frac{\ln\left(\frac{t_{last}}{t_{first}}\right)}{\ln\left(1 + \frac{4R}{\sqrt{N}}\right)} \qquad (4.17)$$

Table 4.5 Column lengths and their application

Column length (m)[a]	Approximate number of compounds in sample	Comments
5	<10 or irrelevant	Wide bore columns for very fast separations where resolution is either easy or unimportant as in simulated distillation (see Chapter 8)
10	10–20	Mostly used for wide bore columns, low to medium resolution applications
25	20–50	The most popular length. Will serve most applications
50	50–100	Should only be used to replace 25–30 m column if resolution is not *quite* sufficient
100	>100	Usually the longest conventional length available commercially. Only required for very complex mixtures.

[a] These are nominal values. Commercial lengths may differ from these marginally, e.g. 60 m in place of 50 m but will not produce significant differences in resolutions (See text).

THE OPEN TUBULAR COLUMN

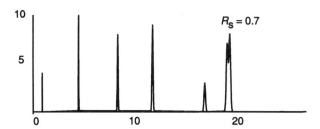
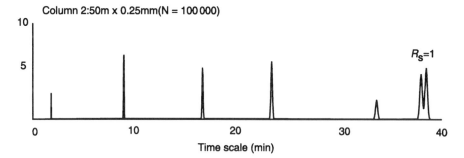

Figure 4.10 Effect of column length on resolution

Where t_{first} and t_{last} are the total retention times of the first and last components eluted from the column and R is the required resolution. This expression is derived on the basis of isothermal operation where, if the plate number N is assumed to remain constant then the peak width increases in direct proportion to the total retention time. However, the value of C_p will not change significantly for programmed temperature operation, because although peak width does not increase at the same rate as in isothermal operation this is qualitatively compensated for by a decreasing resolving power.

What constitutes an acceptable analysis time is entirely a matter for the user. A convenient range, however, for practical purposes is $k = 1 - 20$ as this should ensure that the separations are completed within 60 minutes on most commercial columns. Also, if we specify a minimum resolution of 1.5 then an approximate form of equation 4.17 can be derived as follows:

$$C_p = 0.39 \times \sqrt{N} \tag{4.18}$$

Obviously the components of real samples will not be distributed so conveniently but it is reasonable to take 50% of the peak capacity as guidance in choosing an appropriate column length.

The peak capacity is related to another property known as the **separation number** (SN), or **Trennzahl**. This is used for programmed temperature operation where the simple calculation of plate number is inapplicable and is defined as the maximum number of components that can be resolved to a specified resolution between the peaks of two consecutive *n*-alkanes with z and $(z + 1)$ carbon atoms in their molecules, namely,

$$SN = \frac{t_{R(z+1)} - t_{R(z)}}{W_{h(z)} + W_{h(z+1)}} - 1 \qquad (4.19)$$

4.7.2 General Effects of Column Diameter

The diameter of an open tubular column should always be considered in conjunction with the film thickness as these two parameters have interactive effects on column performance. For example, the theory discussed in Chapter 2, shows that theoretical plate height depends critically on both of these factors and so the **relative** magnitudes of the non-equilibrium contributions, C_g and C_s, cannot be considered from either the diameter or film thickness independently.

The effects of column diameter on permeability show that choosing a column diameter of 0.25 mm gives us a wide choice of column length while still maintaining a low pressure drop across the column. The range of column lengths available commercially for columns with this diameter will realize theoretical plate numbers of 50 000 to 400 000, which is sufficient for most

Table 4.6 Column diameters and typical applications

Column diameter (mm)	Comments
<0.1	Highly specialized research applications and supercritical fluid chromatography. Requires specialized equipment for sample application and detection.
0.1–0.15	Very high speed high resolution applications. Can be used with normal capillary GCs. Available commercially.
0.2–0.35 (0.25 = normal 0.32 = commonly used alternative)	Most commonly used range for general purpose high resolution capillary GC. Wide choice of stationary phases available commercially
>0.5 (0.53 = standard for fused silica wide bore columns)	'Wide diameter' (wide-bore, mega-bore) columns. Low to medium resolution applications, trace analysis, and a faster and better alternative to packed columns.

practical purposes. In the current treatment, columns having diameters of 0.2 mm–0.32 mm will be referred to as **normal bore**, larger diameters will be termed **wide bore**, and smaller diameters than 0.25 mm will be termed **narrow bore**.

Column diameter affects two of the three factors in the resolution equation, namely, plate number and retention factor, but the diameter should not be chosen as the primary way of altering these properties in practice. The choice between either a narrow or wide bore column should depend on such factors as the need to speed up the analysis, to combine the column with a mass spectrometer, to increase the capacity of the column, or to facilitate on-column injection techniques. The recommended features upon which these choices should be based are summarized in Table 4.6.

Narrow Bore Columns

The Golay theory predicts that when $C_g \gg C_s$ the optimum gas velocity increases in inverse proportion to the diameter, and the minimum plate height increases in direct proportion, namely,

$$\bar{u}_{opt} = \frac{D_g}{d_i}\left[\frac{2}{f(k)}\right]^{0.5} \tag{4.20}$$

and

$$H_{min} = 2d_i[2f(k)]^{0.5} \tag{4.21}$$

where

$$f(k) = \frac{(1 + 6k + 11k^2)}{96(1 + k)^2} \tag{4.22}$$

The total separation time is given by the total retention time of the last eluted component, namely,

$$t_R(\text{last}) = \frac{L[1 + k(\text{last})]}{\bar{u}_{opt}} \tag{4.23}$$

where $t_R(\text{last})$ = the total retention time of the last component of the sample separated under optimized conditions in which all the relevant components are resolved by the required amount. L is the column length which is necessary to achieve this resolution for the column diameter chosen, and $\bar{u}$ is the appropriate optimum average carrier gas velocity. Substituting equations (4.20) and (4.21) into (4.23) we obtain the following expression for the analysis time:

$$t_R(\text{last}) = \frac{2d_i^2 N f(k)(1 + k)}{D_g} \tag{4.24}$$

This shows that the necessary analysis time is proportional to the square of the diameter if N, $f(k)$ and D_g are maintained constant. To maintain a constant plate number the column length must be changed in direct proportion to the diameter. Also, to maintain a constant retention factor the film thickness must also be changed to keep its ratio to the diameter constant because

$$k = \frac{K}{\beta} = \frac{4Kd_f}{d_i} \qquad (4.25)$$

Figure 4.11 shows the effect of decreasing column diameter on separation speed in which the retention factor and plate number are maintained constant.

Equation (4.24) assumes both a negligible effect due to gas compressibility and that $C_g \gg C_s$. Very narrow bore open tubular columns, however, require high pressure drops to operate them and both of these factors tend to reduce the predicted gain in separation speed as predicted by this equation. A more realistic relationship is,

$$t_R(\text{last}) = \frac{2d_i^n Nf(k)(1+k)}{D_g} \qquad (4.26)$$

where $n = 1 - 2$

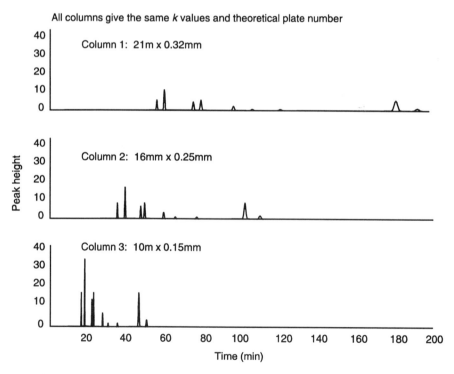

Figure 4.11 Effect of column diameter on separation speed

THE OPEN TUBULAR COLUMN

In view of the compressibility effects at high pressure drops, it is unlikely that columns having smaller diameters than about 0.1 mm will become a useful option for most users. However, they should not be discounted for specialized research applications where very rapid separations may be necessary, as in the study of reaction kinetics or other situations involving short-lived compounds.

An examples of a high speed separation using a 0.1 mm diameter narrow bore columns is shown in Figure 4.12.

Wide bore (WB) columns

The 'standard' diameter for fused silica wide bore columns is 0.53 mm (530 μm), and for metal columns, 0.50 mm. Although the terminology applied to these columns varies between wide bore, large diameter or mega bore, the description wide bore, or WB, will be used herein.

WB columns have become very popular in recent years, being in fact far more ubiquitous than the narrow bore columns described earlier. Thus, while

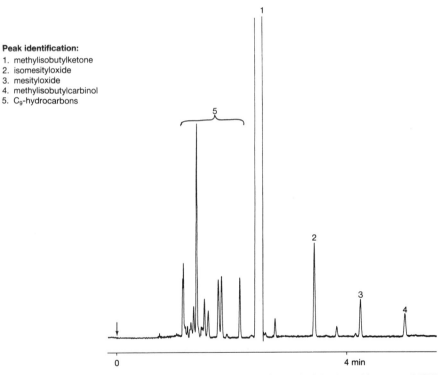

Peak identification:
1. methylisobutylketone
2. isomesityloxide
3. mesityloxide
4. methylisobutylcarbinol
5. C_9-hydrocarbons

Figure 4.12 Example of high speed separation of methyl isobutyl ketone (MIK) and impurities on 0.1 mm diameter column. Column: 10 m x 0.1 mm CP wax 57 CB (0.2 μm); temperature: 50(2 min) to 200 °C at 5 °C min^{-1}; N_2 carrier gas; split injection at 225 °C; detector FID. (Reproduced by permission of Chrompack International BV)

there seems to be little requirement to increase the speed of analysis beyond that already available from normal bore columns, there is a considerable interest in achieving higher capacity, greater detectability for trace analysis and simplicity of operation.

In many ways WB columns fill the gap between the low resolutions associated with conventional packed columns and the extremely high resolutions obtained with 0.25 mm diameter open tubular columns. Separations that do not require the very high column plate numbers of open tubular columns are still performed on packed columns in many laboratories with a prevailing attitude that the analysis is 'safe' because it involves a simpler technique and the results are more reliable. While there was formerly some justification for this attitude, modern columns with their associated techniques and instrumentation have changed this situation and open tubular columns are now widely used in standardized methods of analysis. If high theoretical plate numbers are not necessary to the separation then the use of open tubular columns operated at fast flow rates will usually provide a better alternative to packed columns.

The advantages of the WB column are a consequence of its geometry. Its large internal diameter gives the column a higher permeability than either a packed or a normal bore open tubular column (Table 4.1) and a higher capacity for sample than smaller bore open tubular columns (Table 4.3). Also, referring to (3.7) we see that the effects of extra-column volumes become very much less in comparison with the peak volume variance as the diameter increases. This means that direct injection can be used (Chapter 6) and the specialized facilities normally required for other forms of sample introduction into capillary columns are not required. The combination of high column capacity and the ability to use direct injection techniques means that the columns are particularly suited to trace analysis.

Figure 4.13 shows a typical Golay plot for a WB column, indicating its operating modes. Because of its high permeability, the column can be operated over a very wide range of gas velocities, ranging from optimum to many times the optimum. We can loosely define two extreme operating modes, namely the **packed** and **capillary** mode. The packed mode is the most widely used in practice where the column is operated at high gas velocities. In this mode column efficiency is sacrificed to achieve greater separation speed. Provided the required resolution is maintained, the loss of theoretical plates is not important.

In the packed mode region the longitudinal diffusion contribution to the plate height becomes insignificant, hence

$$H = (C_g + C_s)\bar{u} \qquad (4.27)$$

We see that the plate height increases approximately in proportion to the operating average gas velocity when this is well above the optimum. This

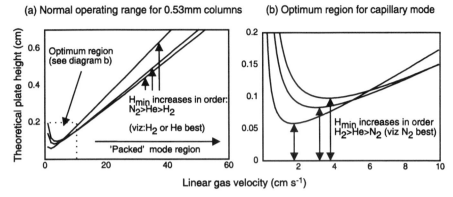

Figure 4.13 Golay plot showing operational features of wide bore columns

illustrates the inherent versatility of the WB column because if the resolution is too low under these conditions then it can be increased by reducing the gas velocity towards the optimum, although of course this will also increase the separation time. The best choice of carrier gas under these conditions is either helium or hydrogen, with helium being the preferable choice for safety reasons. According to (4.27) the choice of carrier gas becomes less important for thick films of stationary phase, i.e. as C_s becomes predominant. Another advantage of the packed mode of operation is that detection limits decrease with mass sensitive detectors such as the FID. This is an additional bonus for trace analysis.

In the capillary mode of operation the gas velocity is set at its optimum value. This invokes a different requirement for the choice of carrier gas according to the Golay theory. Wide bore columns are usually coated with thicker films of stationary phase than smaller bore columns and so C_s will be significant, if not larger than the C_g contribution. As discussed in Section (2.4.5), a heavy carrier gas should be used under these circumstances to produce the maximum number of theoretical plates. This recommendation certainly applies to WB columns coated with a stationary phase film thickness of greater than 2 μm.

Figure 4.14 shows chromatograms from several commercial WB columns operated in the packed and capillary modes.

A summary of the advantages and disadvantages of WB columns operated in these two modes is given in Table 4.7.

In the opinion of some experts the claimed advantages of wide bore columns are unjustified by the scientific facts, i.e. better results could be achieved using shorter normal bore columns operated at optimum conditions. With regard to column performance alone there is some merit to this argument

(a)

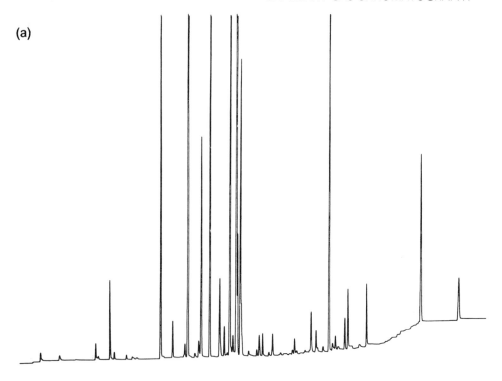

(b)

1. Benzene
2. Toluene
3. Chlorobenzene
4. Ethylbenzene
5. 1,3-Xylene
6. 1,4-Xylene
7. 1,2-Xylene
8. Styrene
9. Isopropylbenzene
10. Bromobenzene
11. n-Propylbenzene
12. o-Chlorotoluene
13. 4-Chlorotoluene
14. 1,3-5-Trimethylbenzene
15. tert-Butylbenzene
16. 1,2,4-Trimethylbenzene
17. 1,3-Dichlorobenzene
18. sec-Butylbenzene
19. 1,4-Dichlorobenzene
20. 4-Isopropyltoluene
21. 1,2-Dichlorobenzene
22. n-Butylbenzene
23. 1,2,4-Trichlorobenzene
24. Naphthalene
25. 1,3-Hexachlorobutadiene
26. 1,2,3-Trichlorobenzene

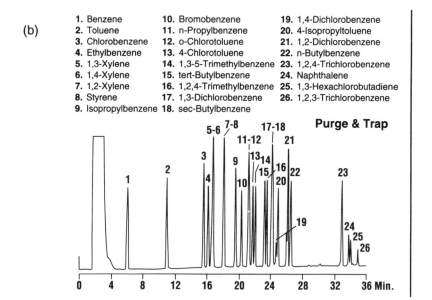

THE OPEN TUBULAR COLUMN

Table 4.7 Advantages and disadvantages of wide bore columns

Property	Advantages	Disadvantages
Plate number	More efficient than packed columns	Less efficient than normal bore columns of similar length
Speed	Faster than packed in 'packed mode'	Slower than normal bore. slower than packed in 'capillary mode'
Capacity	Higher than normal bore	Less than packed
Detectability in trace analysis	Better than normal bore Similar to packed	
Sample introduction	Similar to packed Much easier than normal bore	
Peak shape	Generally better than packed	
Column regeneration	Immobilized phases can be rinsed to remove contamination	More easily contaminated than packed
Permeability	Enables wider range of gas velocities than any other type of commercial column, packed or open tubular	

but the versatility of the WB column would be lost, i.e., the range of operating efficiencies achievable within the pressure range of the equipment. Sensitivity of detection would also decrease with the mass flow sensitive detector as we can see by replacing W_i in (3.14) with W_{max} and substituting this into (4.16). This gives the detector response towards a component when loaded at maximum column capacity:

$$\text{Response} \cong \frac{ad_i^2 \bar{u}\sqrt{\pi}}{8\sqrt{2KL}} \quad (4.28)$$

These disadvantages are further compounded by the more difficult sample injection technique required for smaller diameter columns. For trace analysis, columns should be loaded close to their mass capacity, and while this is a

Figure 4.14 (a) Chromatogram of WB column capillary mode separation of essential oil. Column: 25 m × 0.53 mm FSOT, coated with 1 μm film bonded carbowax; temperature: 60 (1 min) to 225 °C at 8 °C min^{-1} carrier gas H$_2$ at 3 ml min^{-1}. Detector: FID at 270 °C. Injection: split at 250 °C. Run time: 26 min. (Reproduced by permission of Quadrex Corporation.) (b) Chromatogram on WB column packed mode separation of purgeable aromatics. Column: 30 m × 0.53 mm FSOT, coated with 3 μ film DB-624; temperature: 40(5 min) to 135 °C at 3 °C min^{-1}; carrier gas: helium at 8 ml min^{-1}. (Reproduced by permission of J & W Scientific)

comparatively simple matter for wide bore columns using direct injection, it requires either on-column or splitless injection for narrower bore columns (see Chapter 6).

In the author's opinion the WB column gives more choice to the user but the decision to apply it to any particular situation should be guided by a clear understanding of the benefits and the conditions necessary to maximize these benefits. Whether these advantages would increase if the diameter were to be increased beyond 0.53 mm is, however, dubious. The permeability of 0.53 mm columns is already sufficiently high to enable the useful range of operating velocities to be covered without the need for excessive pressure drops. Increasing the diameter further is unnecessary from this point of view and would only result in much longer analysis times. In fact the price paid in analysis time would certainly overtake that associated with packed columns, and the remaining features of increased capacity and delectability would also be better served with a packed column.

4.7.3 Effects of Film Thickness in WCOT Columns

According to the Golay theory the theoretical plate height decreases with increasing film thickness due to the effect of this on the C_s term [(2.48) and (2.49)]. This implies that thin films of stationary phase should be used wherever possible. Why should we ever choose to use a thicker film, and why are such columns commercially available?

As mentioned previously, one of the major disadvantages of the early columns was the poor resolution experienced with very volatile compounds because of insufficient retention by the very small loading of the stationary phase. Attempts to increase the layer thickness to overcome this problem merely resulted in the formation of droplets and film instability. The reason for the poor resolution of volatile compounds is easily understood from the resolution equation as set out in Table 4.4. The relevant factor is $k/(1+k)$ which has a critical effect on resolution for small k values as shown in Figure 4.15 by the plot of this function versus k.

Obviously there can be no resolution when $k = 0$, and the most critical range is $k = 0 - 3$. For $k > 3$ there is only a marginal benefit to be gained by increasing the retention factor further if the effect on the function $k/(1+k)$ is the only effect on resolution. This last comment is of great importance as there are two ways to increase the retention factor with the same stationary phase, i.e., to increase the film thickness d_f or to decrease the temperature. Increasing d_f increases k proportionally, and so the effect on $k/(1+k)$ and consequently on resolution is predictable except for a relatively small counter effect due to the lower theoretical plate number of thicker film columns.

Adjustment of the column temperature is the appropriate way to achieve the required retention range but this can also have an unpredictable effect on

THE OPEN TUBULAR COLUMN

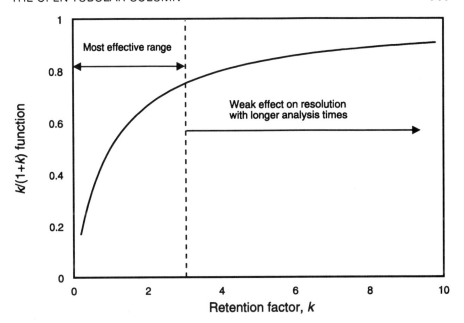

Figure 4.15 Effect of retention factor on resolution

selectivity, as will be discussed in Section 4.9. Reducing the column temperature can either increase or decrease the overall resolution between any two compounds depending on the relative effects on both k and α. Nevertheless the function $k/(1 + k)$ will almost certainly be predominant when $k < 3$ and so reducing the column temperature is the correct action in these cases providing there is sufficient latitude to do this. This action would also extend the overall analysis time and so the use of a lower column temperature for the volatile components may need to be combined with an expedient use of programmed temperature operation to avoid long elution times for the later components. However, if the column temperature is already too low to allow any further reduction then the use of a thicker film of stationary phase is the only alternative.

The early problems associated with the limited retention of volatile compounds were solved in several ways. One of these was to first coat the columns with a layer of porous inert support material and then to apply the stationary phase to impregnate this support, as with conventional packed columns. These column were called **support coated open tubular**, or **SCOT** columns, and were commercially available. They have now become obsolete because thick stationary phase layers can be applied to modern columns by the various processes of immobilization as discussed previously.

Table 4.8 summarizes the range of film thicknesses commonly available commercially, and an indication of choice in relation to the boiling range of

Table 4.8 Choice of film thickness in open tubular columns

$d_f(\mu m)$	min B. Pt (°C)	max B. Pt (°C)	Comments
0.1–0.5	100	500	Most widely used range. Gives highest possible theoretical plate numbers. Applicable to medium to high boiling compounds.
0.5–1.0	50	500	Slightly lower plate numbers but may be chosen for lower boiling compounds or to avoid tailing for polar compounds.
1.0–2.0	30	450	Range usually available for low boiling compounds but with lower plate numbers.
2.0–5.0	ambient	400	Available for some stationary phases for very low boiling compounds, gases, and to avoid tailing for very polar compounds.

the sample. While the main reason for choice is to produce sufficient retention of volatile compounds, another practical reason may be to reduce peak tailing effects with polar compounds. Thus a thick layer of the phase is usually successful in masking these effects to the extent that even very polar compounds can be separated on non-polar phases. Figure 4.16 is an example of this showing the separation of volatile amino-alcohols in dilute aqueous solution.

It should be noted that the choice of film thickness should be made from a consideration of the more volatile components of the sample rather than the high boiling components. Thus if the film thickness is too thin then the retention factors for the volatiles may be too low to allow their resolution at the lowest practicable operating temperature for the column. Provided that the chosen column gives sufficient retention for the low boiling end of the sample then excessively long retention times for the high boiling end can usually be avoided by temperature programming.

4.8 THE STATIONARY PHASE IN CAPILLARY GC

The function of the stationary phase in GC is to selectively retard the components of mixtures sufficiently to produce the required separation. The important properties are retention, as measured by the retention factor k, and the separation factor, α. The separation factor has the most critical effect of the three factors affecting resolution according to the resolution equation. For instance, a change of selectivity from 1.01 to 1.02 will double the

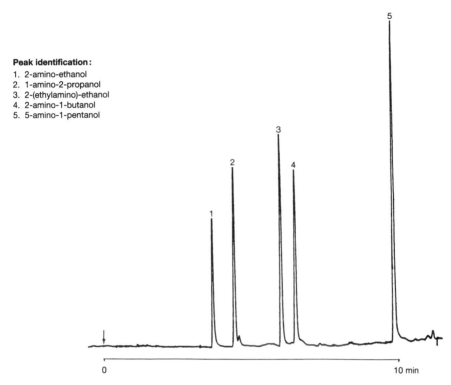

Figure 4.16 Use of thick film column for the prevention of peak tailing of volatile amino alcohols in aqueous solution. Column: 50 m × 0.53 mm FSOT coated with 5 μm layer CPSil5CB(a polymethylsiloxane). Temperature: 65 to 100 °C at 10 °C min^{-1}. Hydrogen carrier gas at 25 cm s^{-1}; split injection; detector FID. Sample: 0.1% approx. of each component in water. (Reproduced by permission of Chrompack International BV)

resolution since R_s is proportional to $(\alpha - 1)$ for closely eluting components. It is not surprising then that the correct choice of stationary phase can be a very important factor for difficult separations. However, the very high theoretical plate numbers of open tubular columns ensures that most components of complex mixtures will separate according to vapour pressure differences on almost any phase. Thus, one of the strengths of the method is the non-critical nature of the phase in many applications and so relatively few phases need to be considered for choice. There are exceptions to this, of course, particularly in the case of mixtures containing isomers or other very similar compounds.

The major problem with very complex mixtures is that while most of the components will separate fairly easily on open tubular columns, there will usually be several critical pairs that are not resolved sufficiently. A typical example is the separation of the polychlorinated biphenyls (PCBs), an important environmental problem requiring accurate monitoring. The analysis is a very complex one since there are 209 possible compounds, many of which are isomers. Figure 4.17 shows a comparison of the separations achieved on two columns containing slightly different stationary phases. On column (a) most of the PCBs are separated except for 149/118, and 132/153/105, but on column (b) these two groups are resolved.

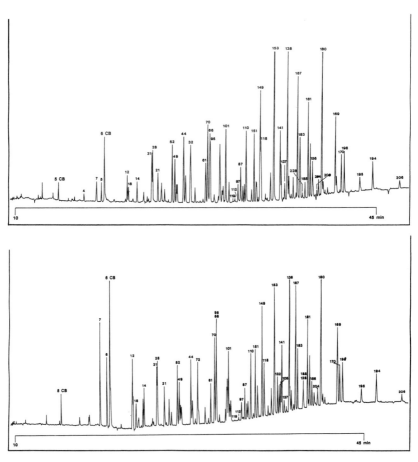

Figure 4.17 Separation of PCBs showing effect of selectivity change. Columns: 50 m x 0.25 mm (0.2 μm) coated with (a) 8% phenyl polymethylsiloxane and (b) 12% phenyl polymethylsiloxane. Temperature: 60 (2 min) to 180 °C (6 min), at 15 °C min^{-1} then at 6 °C min^{-1} to 220 °C (2 min), then 5 °C min^{-1} to 275 °C. Hydrogen carrier gas; detector ECD. Peaks labelled according to the Ballschmitter system for PCBs. (Reproduced by permission Chrompack International BV)

4.8.1 Theory of Retention

The relationship between the amounts of solute in the gas phase and the stationary phase at equilibrium is expressed by the general equation

$$p = \gamma p_0 n_s \tag{4.29}$$

where p and p_0 are the partial and saturation vapour pressures of the solute at the column temperature, γ is the activity coefficient (taken to be equal to the standard state value γ_0 at infinite dilution), and n_s is the mole fraction of the solute in the stationary phase. The general linearity of this relationship is known as Henry's law, and in the specific case of an ideal solution, where γ is unity, the relationship becomes Raoult's law.

The partition coefficient, i.e., the ratio of the **concentrations** of the solute in the two phases is derived as follows.

The ideal gas law states that

$$pv = n_g RT \tag{4.30}$$

where v = volume occupied by n_g moles of the solute, R = the ideal gas constant, and T = column temperature (K).

To obtain an expression for the partition coefficient K we need to express the amounts of the solute in the two phases in mass per unit volume units, namely,

$$\text{Concentration of solute in gas phase (g ml}^{-1}) = \frac{n_g M}{v} \tag{4.31}$$

and

$$\text{Concentration of solute in liquid phase (g ml}^{-1}) = \frac{n_s M \sigma_s}{M_s} \tag{4.32}$$

where M = molecular weight of solute; M_s = molecular weight of stationary phase and σ = density of stationary phase.

Substituting the values of n_g and n_s derived from (4.29) and (4.23) into (4.31) and (4.32), and taking the ratio of the two expressions we obtain the partition coefficient K, namely,

$$K = \frac{\sigma_s RT}{\gamma p_0 M_s} \tag{4.33}$$

The separation factor α for two components A and B is the ratio of the partition coefficients, i.e.,

$$\alpha = \frac{K_A}{K_B} = \frac{\gamma_B p_0(B)}{\gamma_A p_0(A)}$$

$$= \frac{\gamma_B}{\gamma_A} \text{ (interactions)} \times \frac{p_0(B)}{p_0(A)} \text{ (vapour pressure)} \tag{4.34}$$

We now see that selectivity depends partly on the relative vapour pressures of the two components and partly on the ratio of their activity coefficients. In the event of ideal behaviour the activity coefficients are unity and so separation will only occur if the vapour pressures are different. Thus compounds having identical boiling points are unlikely to separate (separation is possible if the two compounds have the same boiling points but different vapour pressures at the column temperature). Fortunately ideality of behaviour is rare and we can usually utilize differences in the affinity of the components for the stationary phase to produce the necessary selectivity.

4.8.2 The Meaning of Polarity in Capillary GC

The term 'polarity' is often used in a very loose sense by chromatographers to indicate the probable strength of interactions between molecules. The activity coefficient is a measure of these interactions between the solute molecules and the stationary phase expressed in non-specific terms. If the separation is to be achieved by means of selective interactions with the stationary phase then this implies that the activity coefficient ratio in (4.34) must differ from unity. Individual activity coefficients are therefore rarely measured in practice as the separation factor alone is sufficient indication of the strength of any selective interactions.

Molecular interaction is based on the *van der Waals* forces of quantum mechanical dispersion, dipole–dipole attraction, and induced dipole interaction. Dispersion interaction is common to all molecules in solution and is not regarded as a contribution to the polarity of the molecule. Dipole–dipole interaction is based on permanent dipole moments and is usually associated with asymmetrical molecules, particularly those with polar groups such as CO, CHO, COOH, NH_2, NH, N, $CONH_2$, OH, SH, Cl, Br and F. Such situations are common and would be expected to make a major contribution to stationary phase selectivity. Induced polarity is a weak force but nevertheless it is probably significant in some situations such as in relation to the delocalized electron distribution of aromatic compounds.

The strongest interactions in capillary GC are those involving hydrogen bond formation. Hydrogen bonds can form when the solute–stationary phase combination creates proton donor–acceptor situations such as the following:

$$[-F \leftarrow H-], [-N \leftarrow H-], [=N \leftarrow H-], [\equiv N \leftarrow H], \text{ or } [=O \leftarrow H-]$$

(4.34a)

Many organic compounds, such as those containing hydroxyl groups, can act amphoterically, providing both donor protons and acceptor sites. Thus hydrogen bonding is commonplace and responsible for many highly specific interactions in capillary GC.

4.8.3 Common Stationary Phases Used with WCOT Columns

As mentioned previously, the availability of a great variety of stationary phases is not necessary for the majority of applications and it is usually sufficient to choose a phase having similar polarity characteristics to the sample to be separated. If the phase polarity is very dissimilar to that of the sample components so as to produce positive deviations from Raoult's law, then the components will elute earlier than expected. This can be used to advantage, as in Figure 4.16, where a thick film column was used to give symmetrical peaks for highly polar compounds using a non-polar phase. However, this could also result in very low k values with a loss of resolution in some cases.

Many of the stationary phases used in capillary GC are based either on polysiloxanes or polyglycols. The basic structure of these two types of phase are as shown:

$$\text{poly(methylsiloxane)}$$
$$\text{Si(CH}_3)_3\text{—[Si(CH)}_2\text{O]}_n\text{—Si(CH}_3)_2$$
$$\text{poly(ethyleneglycol)}$$
$$\text{HO—[CH}_2\text{CH}_2\text{O]}_n\text{—OH}$$

(4.34b)

The methylsiloxane group, $[\text{Si(CH}_3)_2\text{O}]$, is the basic unit for a range of thermally stable polymers varying from low viscosity liquids to silicone gums or silastomer rubbers, depending on the degree of polymerization. These polymers offer a high degree of diffusivity to organic vapours whatever their physical nature, probably as a result of the flexible nature of the Si–O bond, and the openness of the polymer structure. They are, consequently, very effective when used as stationary phases. These materials, and the polyethylene glycols (PEGs) are available under a variety of different commercial names, and are commonly used phases in capillary GC.

The use of polysiloxanes, in particular, has become widespread in view of their thermal stability at high temperatures, and the ease with which more polar groups can be successively substituted for the methyl groups in the polymethylsiloxane structure. The most popular phases, for instance, are made by inserting phenyl, vinyl, trifluoro, cyanopropyl and cyanoethyl groups into the polymer structure. Thus, there is a range of thermally stable polymers for use in capillary GC having polarities ranging from very low to very high. Table 4.9 gives the identity of many of these with some of their popular commercial names.

The substituent groups, with the exception of phenyl, are all proton acceptors and capable of variable amounts of hydrogen bonding. The inclusion of phenyl groups confers minor induced polarity effects which, although weak, can be sufficient to introduce sufficient additional selectivity for many

Table 4.9 Common stationary phases in capillary GC

Number	Stationary phase	Identity	Polarity	Maximum temp. (°C isothermal)
1	OV101, OV1, SP2100, SPB1 HP-1, AT-1, DC200, SE30, CPSIL5CB, 007-1, DB1, BP-1 RTx-1, BP-1	Poly(methylsiloxane)	Low	300–350
2	OV73, SE52, SBP-5, BP-5, 007-2, DB-5, HP-5, RTx-5, CPSIL8CB, AT-5,	95% Dimethyl 5 % phenyl poly(methylsiloxane)	Low	300
3	007-1701, DB-1701, BP-10, AT-1701, CPSIL19CB, SBP-1701, RTx-1701,	86% Dimethyl, 7% phenyl 7% Cyanopropyl polysiloxane	Low/medium	275
4	OV-202, OV-210, OV-215, SP-2401	Poly(methyl-trifluoropropylsiloxane) propylsiloxane)	Low/medium	275

5	DB-225, BP-225, CPSIL43CB, HP-225, OV-225, RTx-225, SP2300, AT-225, 007-225	50% Dimethyl, 25% phenyl, 25% Cyanopropyl polysiloxane	Medium	200
6	007-CW, DB-WAX, Supelcowax-10, HP-20M, Stabilwax, CPWAX52CB, AT-WAX, BP-20	Polyethylene glycol	Medium	250
7	BP-21, DB-23, FFAP, OV351, SP-1000, 007-FFAP, DB-FFAP, Nukol, CPWAX58CB, AT-1000, BP-21	Polyetheleneglycol Nitroterephthalic acid Ester ('free fatty acid phase')	Medium/high	250
8	007-23, DB-23, Sp-2330, RTx-2330, CPSIL88, AT-SILAR, BPX-70, Silar-9CP, Silar-10CP	100% Cyanopropyl	High	225

OV = Ohio Valley Speciality Chemical Co., SP, SPB = Supelco Inc., HP = Hewlett Packard Inc, AT = Alltech, DC = Dow Corning, SE = General Electric, RTx = Restek, CPSILCB = Chrompack DB = J&W, BP = SGE, 007 = Quadrex, Silar = Applied Science

important separations. Polyethylene glycol is also easily derivatized to esters, as, for instance, in the case of FFAP, and these also are commonly used as stationary phases.

4.9 EFFECT OF TEMPERATURE ON RETENTION

The effect of temperature on the solute partition coefficient is given by the van't Hoff relationship:

$$\ln(K) = \frac{H_s}{RT_c} + C \qquad (4.35)$$

or

$$\log(k) = \frac{H_s}{2.3RT_c} + C' - \log(\beta) \qquad (4.36)$$

where H_s = the molar heat of solution of the solute; R = the gas constant; T_s = the column temperature (K), C' and C = integration constants and β = the

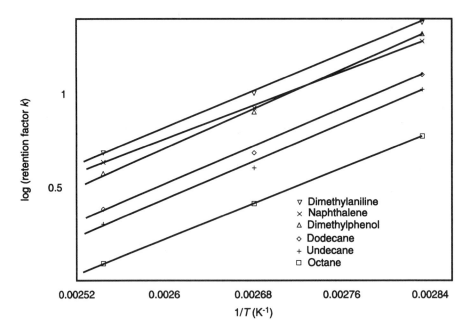

Figure 4.18 Effect of column temperature on selectivity. Column: 25 m × 0.25 mm FSOT coated with 7% phenyl 7% cyanopropyl methyl polysiloxane

column phase ratio. Thus a plot of log (k) versus $1/T_c$ should be linear with a slope of $H_s/2.3R$, provided H_s and C' are constant. Figure 4.18, which gives experimental plots for a number of compounds on a medium polarity phase, shows several interesting features.

The plots are linear over the range of temperatures employed, and furthermore lie approximately parallel to each other indicating that there is little change in selectivity with temperature for many compounds particularly if they are chemically similar to each other. The biggest difference here was for the naphthalene plot which gave a smaller slope than the other components and, consequently, intersected that for dimethylphenol. This means that the peaks for these two compounds will reverse their elution order at about 85 °C.

Figure 4.19 shows the effect of temperature on the resolution of two pairs of compounds selected from the experimental results. The resolution between naphthalene and dimethylaniline declines gradually with increasing temperature due to the effect of the reducing k value on the function $k/(1 + k)$. This is the effect to be expected for most pairs of compounds, and certainly for similar chemical species. The effect on dimethylphenol and naphthalene is more complex as a result of the intersection of the two van't Hoff plots. Thus as the temperature increases from 60 °C the resolution falls rapidly due to the effect of temperature on the selectivity factor, and at about 85 °C, the two

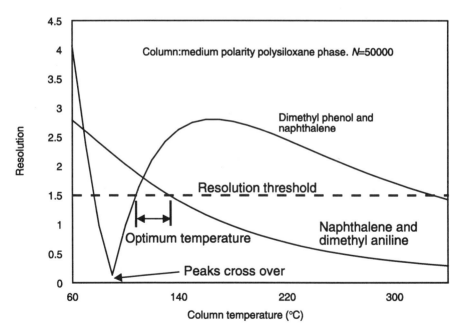

Figure 4.19 Effect of column temperature on resolution (column as in Figure 4.18)

components coelute. Resolution then increases rapidly after the reversal in the elution order, reaching a maximum at about 160 °C. After this the resolution starts to decline because of the reducing retention factor and its effect on the $k/(1+k)$ function. The optimum temperature for the separation of all three compounds is 110 °C.

This kind of behaviour is often experienced when compounds of different chemical type are analysed on moderate to high polarity stationary phases, and it cannot be predicted from runs at a single temperature. **Difficult separations should always include a systematic study at various temperatures to determine optimum conditions for the proposed method.**

The logarithmic relationship between temperature and retention shows that accurate and reproducible control of column temperature is essential. An approximate guide to the magnitude of the effect of column temperature can be derived by replacing the heat of solution in (4.36) with $H_s = E_t \times T_b$ where E_t is the **Trouton constant** and T_b is the boiling point of the component in Kelvin. When H_s is in cal mole^{-1} this constant, expressing the molar entropy of vaporization has a value of 20–23 for many compounds except those having very low molecular weights. Thus assuming that the column temperature is initially at 100 °C, and putting $E_t = 23$, the new column temperature required to halve the retention function k can be derived from,

$$\ln(2) = \frac{23 T_b}{R} \left[\frac{1}{373} - \frac{1}{T_c} \right] \qquad (4.37)$$

Substituting a value of 1.98 cal for R and a boiling point of 373 K for the solute we obtain a new value for the column temperature of 124 °C. Other boiling point and column temperatures produce a similar result, i.e. a change in column temperature of 20 °C will change k by a factor of about 2.

The use of the Trouton approximation in (4.36) also enables a working relationship to be derived between boiling point range and stationary phase film thickness. This provides some guidance on an appropriate choice of this parameter according to the boiling range to be analysed. The recommended film thickness for boiling ranges given earlier in Table 4.8 were derived in this way.

4.10 MEASUREMENT OF RETENTION: THE KOVÁTS RETENTION INDEX SYSTEM

The use of total or adjusted retention times as measurements of retention properties, either for identification or reference purposes, is usually unacceptable because of the relatively large effects of the operating conditions on

THE OPEN TUBULAR COLUMN 153

these properties. It is more satisfactory to use **relative** values to an added reference compound whereby many of the variations are compensated for. Relative retention values are, therefore, widely used within laboratories. However, the agreed standard method for the publication of retention data is the **Kováts retention index system** which is based on a scale defined by the elution of n-alkanes. There is a considerable amount of published data relating to many types of organic compounds [10].

The Gibbs' partial molar free energy of solution is related to the partition coefficient by,

$$-\Delta G_s = RT \ln K \qquad (4.38)$$

The addition of successive methylene groups in a homologous series would be expected to increase the free energy of solution linearly since it only involves the excess energies of transfer to the liquid phase, the creation of a constant incremental cavity volume and additional dispersion interaction with the stationary phase. Thus the excess free energy of solution per CH_2 group should be constant. Also, the retention factor k is proportional to K, and therefore a plot of log k versus the number of carbon atoms for a homologous series should be linear. This has been found to be true for the higher molecular weight members of most homologous series, but does tend to deviate from linearity for the lowest homologues. This linearity is illustrated by the experimental plot given in Figure 4.20.

The Kováts system involves using the linear nature of this plot for n-alkanes as a retention scale for all compounds. Normal alkanes themselves are given an index value of 100 times their carbon number, and the retention index for any other compound is simply its position on this scale as it elutes from the column. In other words it is 100 times the carbon number of a hypothetical n-alkane eluting at the same time and under the same conditions. This is illustrated in Figure 4.20 where the solute of interest elutes between heptane and octane. Thus its retention index must be between 700 and 800, its precise position being obtained by interpolation to the x-axis, as shown, giving a value here of 740. The retention index of solute i can be calculated from

$$I_i = 100C_n + 100\left[\frac{\log t'_R(i) - \log t'_R(n)}{\log t'_R(n+1) - \log t'_R(n)}\right] \qquad (4.39)$$

where C_n = carbon number of last n-alkane (n) eluting before solute i, and n + 1 = first n-alkane eluting after the solute.

This method of measuring solute retention depends only on the solute, the stationary phase and the temperature. Retention indices vary slowly, and almost linearly with temperature and so appropriate corrections can be made by the use of predetermined temperature coefficients.

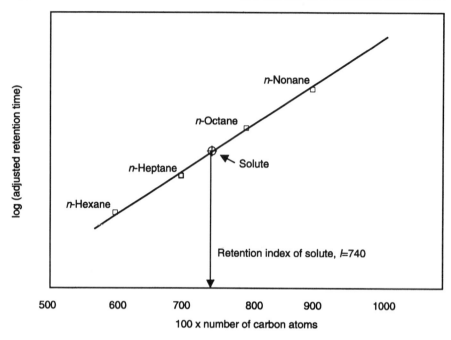

Figure 4.20 The Kováts retention index system. Column: 100 m × 0.25 mm coated with 0.2 μm layer of squalane. Temperature: 70 °C

Some caution should be exercised in measuring Kováts indices on open tubular columns, particularly with thin films of stationary phase. Martin showed that surface adsorption effects can contribute to the overall retention of solutes [11] at low temperatures and on polar phases. This affects the retention of non-polar solutes such as alkanes and so the measurement of retention indices could be in error. Nevertheless there is no doubt that the use of published data can be of considerable help in identification provided that the component type is already known. In the event that the sample type is completely unknown, as may be the case in many environmental situations, then the only satisfactory method of identification is to use an ancillary identification technique such as mass spectroscopy.

An early expectation of the Kováts system was that it would enable retention indices to be predicted by summing group contributions to the molecular structure. We have already seen how this will apply to simple methylene groups in homologous series, namely, that the Kováts index will increase successively by 100 units, but how far can this principle be extended? This approach has met with limited success in practice because of the complex nature of intramolecular interactions when the attached groups are polar.

The retention index itself provides little information about the respective polarities of the solute and stationary phase, but by comparing values measured on **two** stationary phases a direct quantitative measurement of the **relative** polarity can be obtained. This idea formed the basis of the now recognized system which was originally proposed by Rohrschneider [12], but later expanded and modified by McReynolds [13].

The Rohrschneider system is based on two thermodynamic propositions, i.e., that the free energy of solution comprises the sum of the free energy contributions attributable to each type of molecular interaction, and that each of these contributions is a linear combination of the contributions from both the solute and stationary phase. The Gibbs' free energy of solution can thereby be attributed to the separate interactions by,

$$\Delta G_s = \sum_{1}^{n} \Delta G_i \qquad (4.40)$$

where $n =$ the number of specific types of interaction, and free energy contribution from each type, which in turn can be stated in the following form:

$$\Delta G_i = \text{solute coefficient} \times \text{stationary phase interaction} \qquad (4.40a)$$

The Rohrschneider characterization system proposes the use of the difference in the retention indices of any solute on the high molecular weight hydrocarbon, squalane (sq) selected as the standard non-polar phase, to that on the phase of interest (sp) namely,

$$\Delta I = aX + bY + cZ + dU + eS \qquad (4.41)$$

where $\Delta = I_{sp} - I_{sq}$; X, Y, Z, U and S relate to specific types of interaction with the stationary phase, and a, b, c, d and e are solute contributions. Thus, if these values are all known then it should be possible to predict retention data for any solute on any stationary phase. This, indeed, was the initial expectation and the main reason for developing the system.

The phase constants X, Y, Z, U and S were determined by measuring ΔI values at 100 °C for five 'probe' compounds carefully selected to exhibit specific types of interaction. The selected compounds were benzene (dispersion), ethanol (dipole orientation and proton donor), methyl ethyl ketone (orientation and proton acceptor), nitromethane (dipole orientation) and pyridine (proton acceptor and dipole orientation). The solute coefficients a, b, c, d and e were measured by multiple regression analysis on a range of stationary phases for which the phase constants were known.

Although the method was found to be reasonably effective for predicting values to within a few units, little further work appears to have been carried out to refine the technique. Errors of this magnitude would give only a limited indication of selectivity, and when it is close to unity no effective prediction of the resolution would be possible.

Rohrschneider's system was modified and developed by McReynolds and now provides an effective classification scheme for stationary phases, although its original purpose was to show that many of the phases in common use were in fact similar to each other. The number of probes was extended to ten with some of the earlier ones replaced by higher molecular weight compounds. A higher temperature of 120 °C was also advocated. Table 4.10 lists the ten probes used in the Rohrschneider–McReynolds scheme.

Only the **major** types of interactions are given in Table 4.10; other weak interactions would also be present in some cases. Those specifically excluded, however, are given in the third column. Thus, the ten probes cover a wide combination of all the possible interactions in solution. Figure 4.21 shows the principle of the Rohrschneider–McReynolds system.

This diagram shows plots of the n-alkanes and one of the ten probe compounds listed in Table 4.10 on two stationary phases, i.e. squalane as the standard non-polar phase and a more polar phase. The alkanes will elute more quickly on the more polar phase and the slope of the graph is smaller. The solute must be also either non-polar or more polar than the alkanes, and, consequently, will elute relatively later on the alkane scale, i.e. its retention index must be either similar to, or greater than that for the alkanes. Hence the **difference** in its retention index on the two phases is a direct measure of the

Table 4.10 McReynold's probes and their interactions

Probe compound	Main interactions	Comments
Benzene	Dispersion induced polarization by delocalized electrons	No proton donor capability
n-Butanol	Proton donor Proton acceptor Dipole orientation	
2-Pentanone	Dipole orientation Weak proton acceptor	No proton donor capability
Nitropropane	Dipole orientation Weak proton acceptor	
Pyridine	Proton acceptor Dipole orientation	No proton donor capability
2-Methyl 2-pentanol	As butanol	
iodobutane	Dipole orientation	No proton donor or acceptor capability
2-Octyne	As benzene	Different electron mobility
Dioxane	Proton acceptor	No proton donor capability
cis-Hydrindane	Dispersion	No specific interactions possible

THE OPEN TUBULAR COLUMN

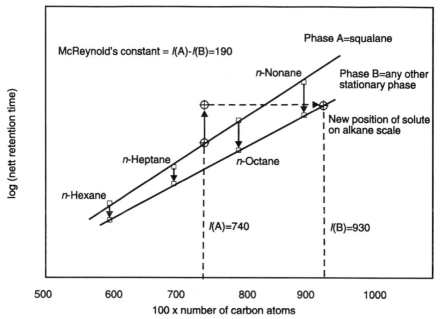

Figure 4.21 McReynolds' constants for stationary phase classification

stationary phase polarity with regard to the specific type of interactions involved.

The McReynolds constant (MC) for the particular probe compound on the stationary phase under study is therefore given by,

$$MC_{sp} = I_{sp} - I_{sq} \tag{4.42}$$

This system has found wide application in gas chromatography and is now the accepted system for stationary phase classification. Typical data is given in Table 4.11 for the stationary phases listed in Table 4.9.

Many manufacturers list McReynolds constants for stationary phases in their catalogues. This can help users to select the most appropriate stationary phases for their purpose. However, as discussed earlier, the very high separation power of modern open tubular columns reduces the need for a meticulous choice to be made but the individual probe constants do give an indication of phase polarity. The major application of the system in modern practice, therefore, is to identify phases, to allocate alternatives, or to enable a phase to be selected which has similar polarity characteristics to the sample.

Table 4.11 McReynolds' constants for stationary phases listed in Table 4.10

Probe	Kováts index on squalane	Squalane	McReynolds constants							
			1	2	3	4	5	6	7	8
Benzene	653	0	16	32	83	146	228	322	340	523
Butanol	590	0	55	72	183	238	369	536	580	757
2-Pentanone	627	0	44	65	141	358	338	368	397	659
Nitropropane	652	0	65	98	231	468	492	572	602	942
Pyridine	699	0	42	67	159	310	386	510	627	801
2-Methyl-2-pentanol	690	0	32	44	141	206	282	387	423	584
1-Iodobutane	818	0	23	23	82	139	226	282	298	480
2-Octyne	841	0	45	36	65	56	150	221	228	298
1:Dioxane	654	0	3	67	138	283	342	434	473	722
cis-Hydrindane	1,006	0	5	9	18	60	117	148	161	267

4.11 SUMMARY OF COLUMN FEATURES

The majority of modern WCOT columns are manufactured from flexible fused silica in lengths of up to 100 m with diameters of 0.1 mm to 0.53 mm, but the **average** column of 25 m × 0.25 mm diameter is suitable for most purposes. Fused silica columns can be used for applications involving column temperatures of up to about 400 °C. Higher temperatures are possible, depending on the choice of stationary phase, but normally require the use of either aluminium clad fused silica or stainless steel columns. Commercial columns are treated during their manufacture to reduce peak tailing effects and some stationary phases can be immobilized inside the column to avoid removal or disturbance by solvents.

Although open tubular columns are intrinsically fast there may be a specific need to increase the speed of analysis. This can be accomplished by choosing columns with smaller than normal diameters (namely less than 0.25 mm). This reduces the length of column necessary for the separation and also increases the optimum gas velocity.

Wide bore columns have become far more ubiquitous than narrow bore columns, particularly in situations where very high resolutions are not necessary. They are easy to use and can replace packed columns for most applications without the need for any extensive modification to the chromatograph. WB columns are also very suited to trace analysis because of their higher capacity and detector response advantages. The extremely high permeability of these columns enables the carrier gas velocity to be adjusted over a very wide range to give either faster analysis at lower resolution, or slower analysis with much higher resolution.

There is a wide choice of stationary phase film thickness available, ranging from about 0.1 μm up to 7 or 8 μm in selected cases. Thin film columns give the highest plate number but may not retain more volatile compounds sufficiently for their resolution and a thicker film should be selected. Thick films can also help to reduce peak tailing if this is a problem with polar compounds.

The precise choice of stationary phase is usually not critical in open tubular columns, although the polarity of the phase should preferably be similar to that of the sample components. For particularly difficult separations the phase should be selected to maximize the chance of specific interactions, particularly those involving proton donor–acceptor effects.

REFERENCES

1. Parker, D. A. and Marshall, J. L. (1978) *Chromatographia*, **11**, 526.
2. Grob, K. (1986) *Making and Manipulating Capillary Columns for Gas Chromatography*, Heuthig, Heidelburg.
3. de Nijs, R. C. M., Rutten, G. A. F. M., Franken, J. J., Dooper, R. P. M. and Rijks, J.A. (1979) *J. High Resoln. Chromatogr.*, **2**, 447.
4. Schomburg, G., Husmann, H. and Borwitzky, H. (1980) *Chromatographia*, **13**, 321.
5. de Nijs, R. C. M., Franken, J. J., Dooper, R. P. M., Rijks, J. A., de Ruwe, H. J. J. M. and Schulting, F, L. (1978) *J.Chromatogr.*, **167**, 231.
6. Carman, P.C. (1956) *The Flow of Gases Through Porous Media*, Butterworths, London.
7. dal Nogare, S. and Juvet, R. S. (1962) *Gas Chromatography–Theory and Practice*, Wiley Interscience, New York, p. 133.
8. Giddings, J. C. (1965) *Gas Chromatography 1964*, (ed. A. Goldup), The Institute of Petroleum, London, pp. 3–24.
9. Keulemans, A. I. M. (1959) *Gas Chromatography*, Reinhold Publishing Corporation, 2nd edn.
10. See *Chromatographic Abstracts*, published for the Chromatographic Society by Elsevier Applied Science.
11. Martin, R. L. (1961) *Anal. Chem.*, **33**, 347.
12. Rohrschneider, L. (1966) *J. Chromatogr.*, **22**, 6. See also Giddings, J. C. and Keller, R. (eds) (1967) *Advances in Chromatography*, vol. 4, M. Dekker, New York, pp. 333–363.
13. McReynolds, W. O. (1970) *J. Chromatogr. Sci.*, **8**, 685.

5

Porous Layer Open Tubular Columns

5.1 INTRODUCTION

A porous layer open tubular (PLOT) column is defined as an open tubular column which is coated on the inner surface with a thick layer of solid porous material. Such columns were first suggested by Golay in 1958 during the discussion on his historical paper in which he introduced the theory of the open tubular column (Chapter 1, reference 27) This theme was later extended when the possibility of using clay-lined columns was advocated as a way of stabilizing the film of liquid stationary phase and of decreasing the stationary phase mass transfer coefficient (C_s) term [1]. However, we have previously seen that one of the problems associated with early wall-coated open tube (WCOT) columns was insufficient retention of volatile compounds. This led to poor separations which were often worse than those previously achieved on simple packed columns with inferior plate numbers. Attempts to rectify the situation by increasing the thickness of the stationary phase film resulted in unstable columns. Surface roughening techniques, as discussed in Chapter 4, help to stabilize the film to some extent but can only have a limited effect. Also, columns which have been subjected to roughening treatments for phase stabilization are not themselves regarded as PLOT columns.

Contrary to Golay's early opinion, equilibrium rates in WCOT thin film columns are limited by the gas phase mass transfer non-equilibrium coefficient (C_g) rather than the C_s term and so now the primary reason for including a porous layer inside the column is to enable the stationary phase loading to be increased to improve the resolution of volatile compounds. Commercial columns of this type were mentioned in Chapter 4 and were indeed marketed as 'support coated open tubular (SCOT) columns'. Golay later modified his theory to accommodate these columns by including configurational factors for the porous layer [2]. Other partition-type PLOT columns designed to increase the liquid phase loading have been described by the author [3] and Nickelly [4].

The development of immobilized phases has now rendered the SCOT column obsolete and so Golay's extensions to his theory must be regarded as irrelevant to modern-day practice. Nevertheless, there has always been an application area in which the use of partition methods lacks success, i.e., in the separation of gaseous and other very volatile compounds, even in cases where there is sufficient retention. **Gas–solid chromatography** has been used extensively and with a great deal of success with packed columns for these situations, and so it is not surprising that the development of analogous open tubular columns has been pursued with a great deal of vigour. The modern PLOT column is invariably based on mechanisms other than simple partition and involves a range of different types of sorbent.

5.2 THEORY OF PLOT COLUMNS

The kinetics of diffusion in PLOT columns are generally more favourable than is the case with many WCOT columns, particularly in relation to the stationary phase non-equilibrium coefficient C_s. This term now depends partly on the rates of diffusion into a thin porous layer, which is essentially a gas phase property with its associated fast rates of diffusion, and partly on the rates of two dimensional surface attachment and detachment to the adsorbent. Thus, even for thick porous layers we should expect the C_s term to be negligible in comparison with the C_g term which should be the same as that for a WCOT column having the same diameter.

The most acceptable theory of PLOT columns is that proposed by Giddings in developing his general non-equilibrium theory of chromatography [5]. According to this theory the C_g term of the Golay equation should be identical for both WCOT and PLOT columns, but a new term C_k is proposed to replace the C_s term, namely,

$$C_k = \frac{8}{c\bar{u}_m} \left(\frac{k}{1+k}\right)^2 \frac{V_g}{S} f_h \tag{5.1}$$

where c = the accommodation coefficient; $\bar{u}_m$ = the mean molecular velocity of solute molecules; V_g = the volume of gas phase in the column; S = surface roughness factor and f_h = the heterogeneity factor. Giddings pointed out that for an open tubular column f_h can be assumed to be unity, c lies in the range 0.1–1, and u_m is roughly equal to the speed of sound. By substituting typical values into this equation it can be shown that

$$C_k \cong 10^{-4} \frac{d_i}{4Sc} \tag{5.2}$$

$$\cong 10^{-4} s - 10^{-6} s \tag{5.3}$$

In these relationships both S and c depend critically on the type of sorbent used. Since C_g is usually in the region of 10^{-3} seconds, $C_k \ll C_g$ unless severe non-uniformity is apparent. Thus PLOT columns should produce high plate numbers which are at least as good as those achieved with WCOT columns. However, this will be dependent on a linear distribution between the gas and stationary phases; a feature which is not normally associated with gas–solid adsorption. Nevertheless, linearity in gas–solid chromatography can be achieved if adsorbent materials are pure and homogeneous in character, and solute concentrations are low. Appropriate modifications may have to be made to the adsorbent in some cases to ensure linearity, as was first shown by Scott, and later in conjunction with Philips, during studies involving the use of alumina adsorbents modified by the addition of various alkali metal salts [6,7]. These treatments were also found to affect selectivity characteristics, an aspect which is now used commercially for the 'fine tuning' of adsorbent selectivity (see Figure 5.3).

Figure 5.1 shows some Golay plots for a 50 m × 0.32 mm i.d. PLOT column coated in this case with aluminium oxide modified by the addition of KCl (see Section 5.4.1).

These plots show several interesting features. As with WCOT columns the minimum plate height is approximately the same for all three carrier gases, indicating that the C_k term **is** insignificant in comparison with the C_g term. Also, H_{min} is approximately equal to the column diameter at the optimum carrier gas velocity, again as predicted by the Golay theory for this situation. There is, however, a considerable difference between the optimum gas velocities experienced here compared with those for WCOT columns. The

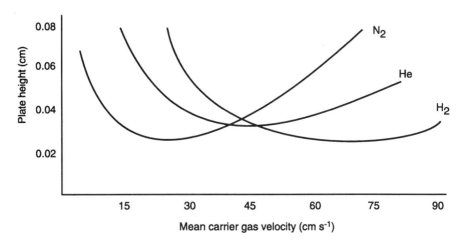

Figure 5.1 Golay plots for a 50 m × 0.32 mm i.d. PLOT column coated with aluminium oxide modified by the addition of KCl

optimum values are now several times faster for the PLOT column. This is quite difficult to explain unless one assumes that the gas phase diffusion coefficient is accelerated in some way by the presence of the porous layer. An intriguing possibility, worthy of further investigation, is that the presence of the porous layer introduces a turbulence effect in the flowing gas stream thereby improving the gas phase mass transfer kinetics. Whatever the theoretical explanation, the practical effect is obviously beneficial with regard to analysis speed.

5.3 THE DEVELOPMENT OF PLOT COLUMNS

There has been a continuing effort to find ways of applying stable and reproducible layers of solid adsorbent materials to the inner surface of capillary tubes. The earliest work was reported by Mohnke and Saffert who successfully separated hydrogen isotopes on etched glass capillaries, and by Kiselev at the same meeting [8,9]. Liberti and his coworkers have also studied etched glass capillaries, preparing SiO_2 layers of 10–50 μm which were used for the separation of organic compounds containing various hydrogen isotopes [10,11]. Other pioneers in this area include Petitjean and Leftault who produced alumina porous layers in aluminium capillaries [12], and Jentzsch and Hövermann who made PLOT columns coated with copper oxide from copper capillaries [13].

All of these early methods involved some form of surface etching to produce an adsorbent layer and although successful in the context of their individual research applications, they did not receive widespread interest. An essential requirement for any commercial column is a high degree of reproducibility coupled with reasonable manufacturing costs. Adsorption is a notoriously variable process and even small differences in the column selectivity would be unacceptable for general use. More promising was the work of Kirkland who prepared adsorption columns by dynamically coating stainless steel capillaries with colloidal boehmite, a form of alumina manufactured by Du Pont. The columns diameters were in the range 0.25–0.5 mm but with lengths of only 7.5 m, and these were used successfully for the rapid separation of volatile fluorinated hydrocarbons [14].

Static coating methods for PLOT columns followed from the early work of Halász and Horváth [15,16]. The principle of static coating is essentially the same as for the manufacture of WCOT columns but the technology is markedly different. Thus, a stable suspension must be achieved with the adsorbent in a colloidal state, and the suspension medium must be removed so as to leave a stable, uniform layer of reproducible characteristics and geometry.

Table 5.1 Types of commercial PLOT column

Porous layer	Temp. range (degs/C)	Typical supplier	Typical column dimensions L, ID (d_f) (where known)	Applications
Aluminium oxide with various additives	−60 to 200	PLOT Alumina™ (Chrompack) AT-Alumina™ (Alltech) GS-Alumina™ (J & W)	10 m–50 m, 0.32 mm (5) and 0.53 mm (10) 30 m, 0.53 mm 30 m and 50 m, 0.53 mm (9)	Light hydrocarbons Naphthas Natural gas Separation of all C1–C4 hydrocarbons Halocarbons
Molecular sieve 5A	−60 to 300	Molesieve 5A™ (Chrompack) AT-Molesieve™ (Alltech) GS-Molesieve™ (J & W) PLT-5A™ (Quadrex)	10–25 m, 0.32 mm (30) and 0.53 mm (50) 30 m, 0.53 mm 30 m, 0.53 mm 15–25 m, 0.32 mm and 0.53 mm	Permanent non-polar gases Deuterated hydrocarbons
Carbon based	As required	CarboPLOT P7™ (Chrompack)	10–25 m, 0.53 mm (25)	Separation of CO and CO_2 from air. Baseline separation of O_2 and N_2. Also C1 and C2 hydrocarbons
Porous polymers	−60 to 250	PoraPLOT Q, S, and U (TM Chrompack)	10–50 m, 0.32 mm (10) and 0.53 mm (20)	Wide range of polar and non-polar compounds. Polar gases (CO_2, nitrogen oxides, water, sulphur gases, CFCs), ketones, esters, acids

POROUS LAYER OPEN TUBULAR COLUMNS

5.4 TYPES OF PLOT COLUMN

Most of the major suppliers of open tubular columns and accessories also supply a range of PLOT column to cover many diverse applications. Table 5.1 summarizes the major types of column and their application field. Unfortunately, much of the detail relating to the methods of preparation and the character of the sorbents is unavailable in view of the highly competitive nature between commercial column manufacturers. Methods of coating are based either on static or dynamic procedures, but are developed to ensure a sufficient degree of reproducibility and performance to satisfy the user. Particle size is 1–5 μm in diameter and layer thicknesses range from 5 to 50 μm depending on the nature of the adsorbent and the diameter of the column.

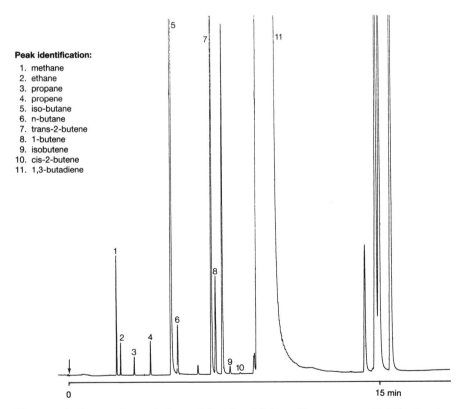

Peak identification:
1. methane
2. ethane
3. propane
4. propene
5. iso-butane
6. n-butane
7. trans-2-butene
8. 1-butene
9. isobutene
10. cis-2-butene
11. 1,3-butadiene

Figure 5.2 Separation of impurities in 1,3-butadiene on a PLOT alumina column. Column: 50 m × 0.32 mm FSOT column coated with 5 μm layer of Al_2O_3 modified by KCl. Temperature: 100 to 200 °C at 6 °C min^{-1}; N_2 carrier gas at 35 cm s^{-1}. Split injection; detector FID. sample size: 100 μl (gas). (Reproduced by permission Chrompack International BV)

5.4.1 Aluminium Oxide Columns

As was mentioned earlier, PLOT columns coated with layers of aluminium oxide have been studied by several schools of workers using a variety of techniques. Schneider and coworkers have also reported the preparation of glass columns coated with alumina using a thixotropic suspension of Al_2O_3 in water [17]. This work was extended by de Nijs and de Zeeuw who described columns coated statically in a similar way but partly deactivated by the addition of potassium chloride [18,19]. These columns form the basis of the commercial columns available today with alternative additives to give required variations of selectivity.

Figure 5.2 shows a typical application of this type of column showing the separation of impurities in 1,3 butadiene. Note the selective nature of the column towards the C4 isomers.

Substitution of the potassium chloride additive by sodium sulphate produces an interesting change in the alkene–alkyne selectivity, as illustrated in Figure 5.3.

5.4.2 Molecular Sieve 5A columns

Molecular sieves are alkali metal alumino-silicates containing various cations in the lattice structure. They separate small molecules on the basis of molecular diameter and ease of diffusion through the micropores. Molecular sieve 5A is the most widely used material of this type, and indeed has been used for many years in packed columns to separate gases such as hydrogen, helium, argon, oxygen and nitrogen. Molecules larger than the micropore size are strongly retained, often irreversibly if there is the likelihood of interaction with the strongly polar nature of the material. Thus polar gases such as carbon dioxide, water and sulphur gases will be very strongly retained by the packing and cannot be determined.

An example of the use of a molecular sieve PLOT column is given in Figure 5.4.

This shows clearly the wide separation achieved between oxygen and nitrogen. Permanent gases of this type cannot be detected by the FID and so a micro-TCD is used instead. In practice, a wide bore column of this type could be used with direct injection from a gas sample valve provided that the sample volume was no greater than about 200 ml.

5.4.3 Carbon Layer Columns

There is little information about the precise nature of columns coated with various types of carbon. Obviously, there are numerous possibilities with regard to the use of carbon in its various forms, but the number of PLOT

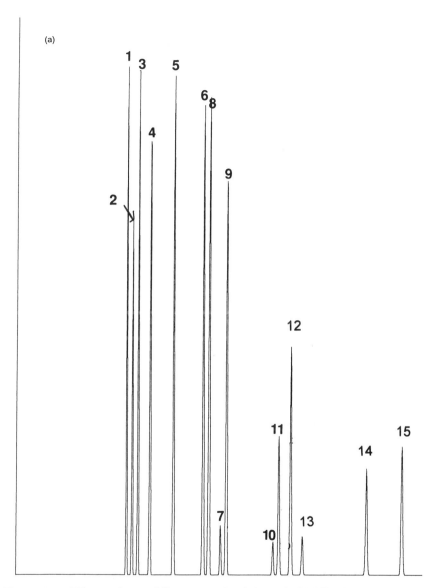

Figure 5.3 Effect of modifier on selectivity of PLOT alumina columns. Columns: 50 m × 0.32 mm FSOT, (a) Al_2O_3/KCl, and (b) Al_2O_3/Na_2SO_4. Temperature: 70 to 200 °C at 3 °C min^{-1}. Nitrogen carrier gas. Peak 1 = methane; 2 = ethane; 3 = ethylene; 4 = propane; 5 = propylene; 6 = acetylene; 7 = propadiene; 8 = isobutane; 9 = *n*-butane; 10 = *trans*-2-butene; 11 = 1-butene; 12 = isobutene; 13 = *cis*-2-butene; 14 = methylacetylene; 15 = 1,3-butadiene. (Reproduced by permission Chrompack International BV)

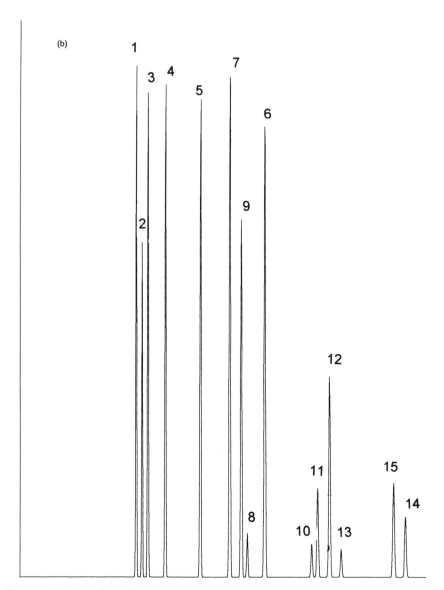

Figure 5.3 (*continued*)

columns available at the present time is limited. The Alltech Associates Carbograph™ columns are coated with a mixture of graphitized carbon black and stationary phase and give useful separations of both polar and non-polar compounds. It is doubtful, however, if these columns can be considered as true PLOT columns according to their definition. The Chrompack 'CarboPLOT P7'

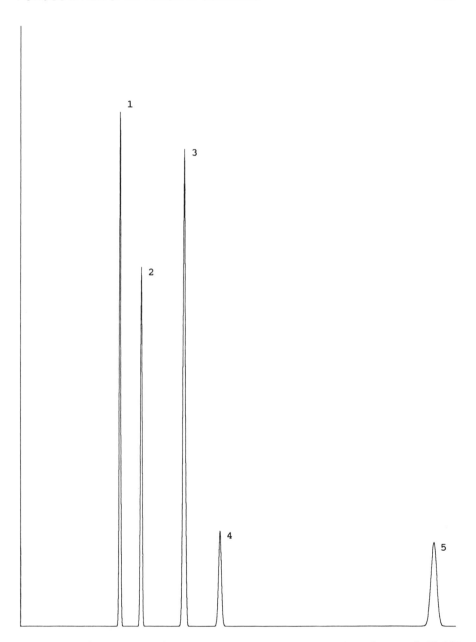

Figure 5.4 Separation of permanent gases on molecular sieve 5 A PLOT column. Column: AT-Mole Sieve, 30 m × 0.53 mm. Temperature: 50 °C; helium carrier gas at 4.6 ml min^{-1}. Peaks: 1, hydrogen; 2, oxygen; 3, nitrogen; 4, methane; 5, carbon monoxide. (Reproduced by permission of Alltech Associates Applied Science Ltd)

column is a true PLOT column, but again little information is available with regard to the nature of the adsorbent which is described as being equivalent to Carbosphere, a type of molecular sieve with a pore diameter 13 Å. Its main field of application is in the separation of permanent gases, including the more polar gases such as CO and CO_2. A useful feature is that, unlike many adsorbents, the column is unaffected by moderate amounts of water in the sample. Figure 5.5 shows a typical use of this type of column for the separation of gases.

5.4.4 Porous Polymer Columns

Porous polymers have been in use in packed columns since the early days of gas chromatography. Their use is varied, having been applied to a wide range of sample types, including permanent gases, polar compounds such as alcohols, amines, carboxylic acids, and non-polar compounds such as hydrocarbons. The commercial polymers are generally described as the 'Chromosorb 100' series from the Johns Mansville Corporation, 'Hayeseps'

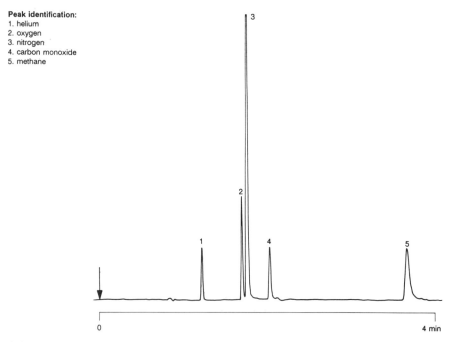

Figure 5.5 Separation of permanent gases on carboPLOT column. Column: 25 m × 0.53 mm FSOT carboPLOT P7 coated with 25 μm layer. Temperature 40 °C. H_2 carrier gas at 48 cm s^{-1}. Split injection; microTCD detector. Sample size: 30 μl. (Reproduced by permission of Chrompack International BV)

POROUS LAYER OPEN TUBULAR COLUMNS

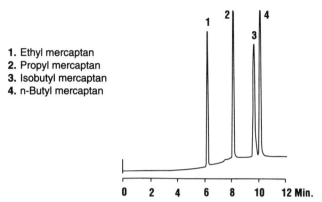

1. Ethyl mercaptan
2. Propyl mercaptan
3. Isobutyl mercaptan
4. n-Butyl mercaptan

Column:	30m x 0.53mm ID
Coating:	GS-Q
Temp:	105°C (1min) to 233°C (3min) at 15°C/min (FID)
Flowrate:	Helium, 8 mL/min

Figure 5.6 Separation of mercaptans on porous polymer PLOT column. Column: 30 m × 0.53 mm GS-Q. Temperature 105 (1 min) to 233 °C (3 min) at 15 °C min^{-1} helium carrier gas at 8 ml min^{-1}. Components: 1 = ethyl mercaptan; 2 = propyl mercaptan; 3 = isobutyl mercaptan; 4 = n-butyl mercaptan. (Reproduced by permission of J & W Scientific)

Table 5.2 Porous polymers used in PLOT columns

Commercial name	Description	Typical applications
PoraPLOT Q[a]	styrene–divinylbenzene	Alcohols, water, polar volatiles, hydrocarbons, gases
PoraPLOT S[a]	divinylbenzene–vinylpyridine	Ketones, esters, halogenated compounds, hydrocarbons
PoraPLOT U[a]	divinylbenzene–ethyleneglycol–dimethylacrylate	Polar volatiles, nitriles, nitro-compounds, alcohols, aldehydes, ethane, ethylene, sulphur gases, CO_2 in air
PoraPLOT amines[a]	(no information)	Ammonia, primary, secondary and tertiary amines

[a] Tradenames of Chrompack International BV

172 CAPILLARY GAS CHROMATOGRAPHY

Peak identification:
1. nitrogen
2. methane
3. carbon dioxide
4. ethane
5. water
6. propane

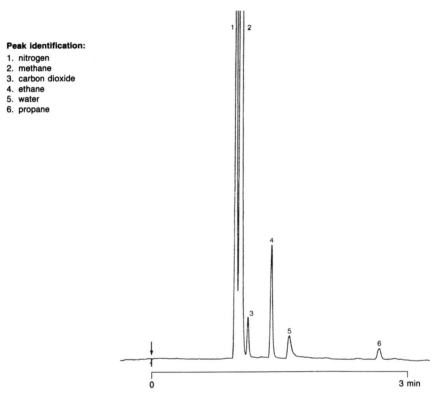

Figure 5.7 Separation of C1–C3 hydrocarbons, CO_2 and water on a porous polymer PLOT column. Column: 10 m x 0.32 mm FSOT poraPLOT Q (10 μm). Temperature: 100 °C. Hydrogen carrier gas; split injection; MicroTCD detection. (Reproduced by permission of Chrompack International BV)

from the Hayes Corporation, or 'Porapaks' from the Millipore Corporation. PLOT columns coated with porous polymers have only become available in recent years following the earlier work of Hollis [20], but have already become established for numerous applications.

The commonest porous polymer in general use is the PLOT equivalent of Porapak™ Q, or Chromosorb 102 and is based on polystyrene crosslinked by divinylbenzene in a two-stage polymerization process. This material is extremely non-polar but, nevertheless, can be applied to a wide range of compounds, both polar and non-polar. An interesting application, for instance, is the determination of water in such matrices as solvents, drug formulations, gases, etc. Figure 5.6 shows the use of a J & W column for the separation of mercaptans.

Figure 5.7 shows the use of the Chrompack PoraPLOT Q column for a wet gas mixture. The water elutes from the column to give a reasonably symmetrical

POROUS LAYER OPEN TUBULAR COLUMNS **173**

peak which can be measured quantitatively, provided that a suitable detector such as a micro-TCD is employed.

Numerous porous polymers are available for packed columns but only a few of these have been incorporated into PLOT columns, probably because there is less need to have the same range of selectivity in view of their much higher plate numbers. Four types of porous polymer column are described in Table 5.2.

The effects of increasing polymer polarity are shown by comparing the three chromatograms given in Figure 5.8 for C1-C3 hydrocarbons. The retention times of the unsaturated compounds is increased in relation to those for saturated compounds, the largest effect being shown by the alkyne triple bond.

There is no information available about the most polar column listed in Table 5.2 except that is a porous polymer with basic groups. The column is suitable for extremely polar amines, including ammonia, as shown in Figure 5.9.

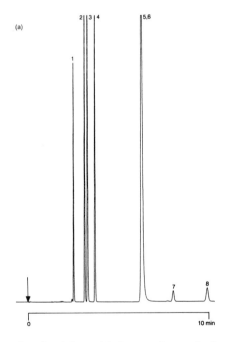

Figure 5.8 Change of selectivity with increasing polarity in porous polymer PLOT columns. Columns 25 m × 0.53 mm (a) poraPLOT Q, (b) poraPLOT S and (c) poraPLOT U. Temperature 80 °C; H_2 carrier gas; split injection; FID. Components: 1 = methane; 2 = ethylene; 3 = ethane; 4 = acetylene; 5 = propylene; 6 = propane; 7 = propadiene; 8 = propyne. (Reproduced by permission of Chrompack International BV)

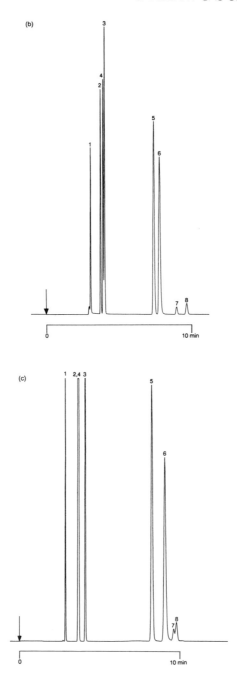

Figure 5.8 (*continued*)

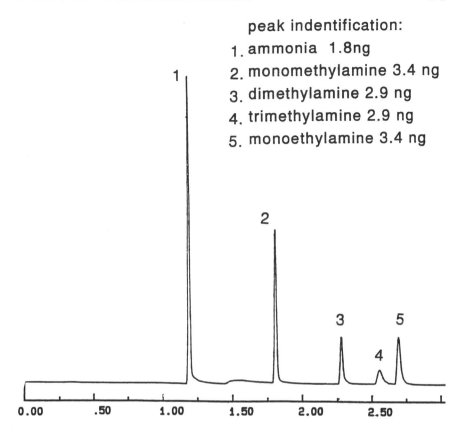

Figure 5.9 Separation of ammonia and volatile amines on porous polymer column modified for amines analysis. Column: 25 m × 0.32 mm poraPLOT amines. Temperature: 140 (2 min) to 250 °C at 10 °C min^{-1}. (Reproduced by permission of Chrompack International BV)

5.5 SUMMARY

For many years numerous forms of PLOT column were described in the literature, but it is only in recent years that they have become accepted as alternatives to packed adsorption columns. This acceptance is largely a result of improvements in manufacturing techniques which has led to a much higher degree of reproducibility and reliability. In view of the number of possible adsorbents that could be used there is no doubt that we can look forward to further developments in this area in due course.

REFERENCES

1. Golay, M. J. E. (1960) in *Gas Chromatography 1960*, (ed. R. P. W. Scott), Butterworths, London, pp. 139–143.
2. Golay, M. J. E. (1968) *Anal. Chem.*, **40**, 382.
3. Grant, D. W. (1971) *Gas–Liquid Chromatography*, van Nostrand-Reinhold, New York, pp. 84–91.
4. Nickelly, J. G. (1972) *Anal. Chem.*, **44**, 623.
5. Giddings, J. C. (1964) *Anal. Chem.*, **36**, 1170.
6. Scott, C. G. (1962) in *Gas Chromatography 1962*, (ed. M. van Swaay), Butterworths, London, pp. 36–66.
7. Phillips, C. S. G and Scott, C. G. (1968) in *Progress in Chromatography*, (ed. J. H. Purnell), Wiley, New York, p. 121.
8. Mohnke, M. and Saffert, W. (1962) in *Gas Chromatography 1962*, (ed. M. van Swaay), Butterworths, London, pp. 216–224.
9. Kiselev, A.V. (1962) in *Gas Chromatography 1962*, (ed. M. van Swaay), Butterworths, London, pp. xxxiv–liii.
10. Liberti, A. (1967) in *Gas Chromatography 1966*, (ed. A. B. Littlewood), Institute of Petroleum, London, pp. 95–113.
11. Bruner, F., Cartoni, G. P. and Possanzini, M. (1969) *Anal. Chem.*, **41**, 1122.
12. Petitjean, D. L. and Leftault, C. J. (1963) *J. Gas Chromatogr.*, **1**(3), 18.
13. Jentzsch, D. and Hovermann, W. (1963) *14th Pittsburgh Conference on Analytical Chemistry and Applied Spectroscopy*, Pittsburgh.
14. Kirkland, J. J. (1963) *Anal. Chem.*, **35**, 1295.
15. Halász, I. and Horváth, C. (1963) *Nature (London)*, **197**, 71.
16. Halász, I. and Horváth, C. (1963) *Anal. Chem.*, **35**, 499.
17. Schneider, W., Frohne, J. C. and Bruderreck, H. (1978) *J. Chromatogr.*, **155**, 311.
18. de Nijs, R. C. M. and de Zeeuw, J. (1983) *J. Chromatogr.*, **279**, 41.
19. de Zeeuw, J. *et al.* (1973) *J. Chromatogr. Sci.*, **25**, 71.
20. Hollis, O. L. (1973) *J. Chromatogr. Sci.*, **11**, 335.

6
Sample Introduction

6.1 INTRODUCTION

The process of sample introduction in GC became known as **injection** when it was found that the use of microlitre syringes was the most convenient way for the sample to access the column without disturbing the flow rate or other conditions. However, the extremely small scale of the open tubular column imposes stringent demands on the quality of the sample and the way that it is injected. Also, any non-volatile material such as decomposition products from thermally labile compounds, or particulates already present, will accumulate at the column entrance and cause a rapid deterioration of performance. Appropriate sample preparation techniques should be applied wherever possible to eliminate these materials and to ensure that trace components are present in sufficient concentration for their satisfactory detection and measurement. These techniques include methods for collecting gas and vapour samples from sample matrices which are unsuitable for direct analysis by capillary GC. Methods of this type come under the general heading of **headspace analysis** where enrichment techniques are often included to increase the component concentrations so that they are above the minimum detectable levels of the GC. In many applications both the sample preparation and injection can be integrated into a single operation by using appropriate equipment, and these techniques will be discussed in Chapter 7. This chapter will only consider the introduction of samples that are already in a satisfactory state for analysis by capillary GC.

Methods of sample introduction in capillary GC have been discussed in books by Lee and coworkers [1], Jennings [2], Sandra [3] and Grob [4,5].

6.2 THEORY OF SAMPLE INTRODUCTION

The objectives of effective sample introduction can be summarized as follows.

- To introduce the required quantity of sample without overloading the column. The 'required quantity' in this context is the amount of sample that admits sufficient masses of its components for their detection and measurement.
- To ensure that the sample entering the column is representative of the original.
- To ensure that the method of sample introduction does not lower the performance of the column in any way, i.e. the column efficiency and retention properties are maintained, and the peaks shapes are Gaussian.

In Chapter 3, we saw how column installation and instrument features could contribute to solute zone dispersion and lower the theoretical plate number. Modern equipment and correct column installation and operation will minimize these extra-column effects but equation (3.2) shows that the efficiency of sample introduction is still a major factor and this is illustrated by Figure 6.1.

The shaded portions show the respective lengths of the column, w_i and w_t, which are occupied by a pure solute at the inlet, immediately following injection, and the outlet, immediately prior to its elution from the column. These lengths are defined by the plate theory as being equivalent to four standard deviations of the Gaussian concentration profiles at these two extreme positions. The final zone width w_t is larger than w_i because of the column dispersion processes described in Chapter 2. Restating equation (3.1) in terms of zone occupancy, and assuming other extra-column effects are negligible:

$$w_t^2 = w_c^2 + w_i^2 \qquad (6.1)$$

where w_c is the column contribution to zone dispersion.

The length of the injection zone w_i depends on the sample injection technique and the column operating conditions, and it is essential to ensure that

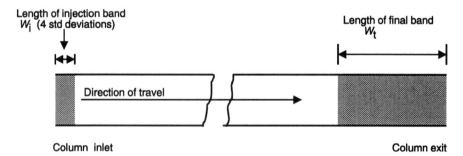

Figure 6.1 Zone occupation at the inlet and exit positions of the column

SAMPLE INTRODUCTION

w_i is much smaller than w_c. A further complication which can arise with a poor injection technique is that the sample vapour enters the column as a tailing distribution rather than as a discrete 'plug', usually as a result of vapour slowly diffusing from semi-stagnant pockets in the injector which become filled during the vaporization stage. Excessive sample size is another common cause of this effect in vaporization injectors because of the rapid formation of a very large volume of vapour which, if it exceeds the geometrical volume of the injector, will back-diffuse into the inlet gas lines of the instrument. As the major component is usually solvent then this shows the biggest effect, i.e., excessive tailing of the solvent peak thus tending to obscure components which elute close to it as illustrated by Figure 6.2.

6.2.1 The Sample Volume Capacity

The volume capacity of the column previously referred to in Chapter 4 is defined as the maximum **volume** of sample **vapour** that can be admitted to the injector before the column performance deteriorates to an unacceptable level. This volume determines the feed time to the column, and hence the initial zone occupancy w_i, although the latter also depends on the injector geometry and the

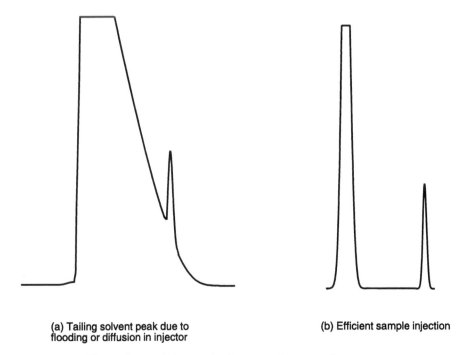

(a) Tailing solvent peak due to flooding or diffusion in injector

(b) Efficient sample injection

Figure 6.2 Effect of poor injector design on solvent peak

operating conditions. Thus, calculation of w_i and its effect on w_t from (6.1) may show that the column can accept a certain volume of sample vapour to give an insignificant occupation of the column but it may still give some peak distortion due to overloading of the injector as noted earlier. It is essential for the operator to understand this limitation since injector overloading will contaminate the system and cause **memory effects** in which peaks from previous samples appear repetitiously in subsequent chromatograms.

Let us consider the effects of the sample vapour volume and the column conditions on the final zone width (w_t in Figure 6.1). It is assumed that the sample is vaporized to produce an effective volume v_{inj} at the column inlet pressure after mixing with the carrier gas. The total sample entry time is then the time taken to purge the injector of its mixture of sample vapour and carrier gas.

If v_{inj} = effective injection volume (ml) and F_c = mean volumetric flow rate through column (ml s^{-}), then

$$\text{time to enter column(s)} = \frac{V_{inj}}{F_c} \qquad (6.2)$$

Since the zone travels at $1/(1+k)$ times the velocity of the carrier gas, then assuming that the injected vapour plug has a discrete Gaussian shape and w_i is equivalent to four standard deviations, then

$$\text{length of injection band } w_i = \frac{\bar{u}V_{inj}}{(1+k)F_c} \qquad (6.3)$$

According to (6.1)

$$\text{final band length } w_t = \sqrt{w_{inj}^2 + w_{col}^2} \qquad (6.4)$$

But, according to the plate theory

$$w_c = \frac{4L}{\sqrt{N_c}} \qquad (6.5)$$

where N_c is the true theoretical plate number of the column (i.e., when there are no extra-column effects). Substituting (6.3) and (6.5) into (6.4)

$$W_t = \sqrt{\frac{\bar{u}^2 v_{inj}^2}{(1+k)^2 F_c^2} + \frac{16L^2}{N_c}} \qquad (6.6)$$

Thus the actual plate number, N_t is $16 \times (L/w_t)^2$ and therefore

$$N_t = \frac{N_c}{\left[\dfrac{N_c v_{inj}^2}{16 v_{col}^2 (1+k)^2} + 1\right]} \qquad (6.7)$$

SAMPLE INTRODUCTION

Defining column volume capacity as the maximum volume of sample vapour, including carrier gas that is formed in the injection space before the column looses 20% of its efficiency, then the following expression can be derived from (6.7)

$$v_{inj}(\max) \leqslant \frac{2v_{col}(1+k)}{\sqrt{N_c}} \qquad (6.8)$$

where v_{col} = geometrical volume of the column.

Hence the volume capacity of the column depends on the column volume, the intrinsic column efficiency and the retention factor. The effect of the retention factor, k, is particularly significant as this can be increased dramatically by lowering the column temperature at the time of injection. We saw in Chapter 4 how lowering the column temperature by 20 °C approximately doubles the retention factor, and so if the column temperature is 100 °C lower than the elution temperature then the effective volume capacity is increased by a factor of 2^5, i.e., about 30 times. Temperature programming, therefore, is a very useful way of increasing the volume capacity of the column, in addition to its main function of extending the analytical range. Indeed it becomes an essential feature of the techniques of splitless and on-column injection as discussed in Sections 6.4.3 and 6.4.5.

As mentioned earlier, one of the main problems with vaporization injectors is that the volume capacity of the column may be larger than the injector is capable of handling because of its geometrical limitations. Practical experience shows that most injectors cannot handle actual sample volumes much larger than the volume of the injector itself without flooding and contamination effects. Thus assuming that the volume capacity is limited by injector volume then this can be used in (6.8) as the value of $v_{inj}(\max)$.

The consequences of (6.7) and (6.8) on injection techniques will be discussed in the appropriate sections.

6.3 INTRODUCING THE SAMPLE TO THE INJECTOR

6.3.1 Liquid Samples

Most samples for GC are prepared as dilute solutions in volatile solvents. These are then introduced to the injector by means of a microlitre volume syringe containing the appropriate volume (usually 0.1–1 μl) of the solution. Injection is normally carried out by piercing a self-sealing septum of thermally stable silicone rubber.

Manual sample injection is undoubtedly one of the biggest causes of analytical error in practice, particularly where different operators are involved. Operator error can be minimized by using one of the many types of commercially available autosampler which are capable of handling either multiple or individual samples according to their individual design. Whether or not autosamplers are used, there are inherent problems associated with the injection of liquid samples into hot injectors and the operator should understand the nature of these problems in order to attain satisfactory analytical standards.

Any injector which involves the conversion of the sample to its vapour before it reaches the column is known as a **vaporization injector**. The vaporization process is usually carried out instantaneously as a **flash** vaporization and this necessitates a rapid and immediate contact between the syringe needle and a very hot zone. If sample is present inside the needle then it would start to vaporize immediately, even before the plunger is depressed. Although this is not a problem for most samples, very high boiling components may either distil slowly from the needle, or they may vaporize and recondense in its cooler upper regions. In either case the sample may be partially retained by the syringe when it is withdrawn from the injector and this will cause **discrimination errors**. However, as noted in Chapter 3, modern injectors can be operated at temperatures of more than 400 °C, and this should be sufficient for samples with boiling points of up to at least 500 °C. On-column injection, which does not involve initial vaporization of the sample, is nevertheless recommended wherever possible and particularly for high boiling samples. Users should beware of using very high injector temperatures which may cause pyrolysis of the sample.

Capillary GC requires minute sample volumes in the microlitre or submicrolitre range, and specially designed syringes are available for the purpose.

6.3.2 Injection Syringes and Syringe Handling

Two basic types of microlitre syringe, known as **plunger in barrel** (PIB) and **plunger in needle** (PIN), are used. PIB syringes have a conventional design in which the sample is withdrawn and expelled from a calibrated glass barrel by operating the plunger. These syringes are available with capacities down to 5 μl from which it is possible to inject quantities down to about 1 μl, i.e., sufficiently small for capillary operation.

PIN syringes have a similar appearance to the PIB type but now the glass barrel and visible part of the plunger is only for measurement and handling purposes as the entire sample is contained in the syringe needle, as illustrated in Figure 6.3.

The lower part of the plunger consists of a fine wire which slides inside the needle, and a Teflon™ seal ensures that a partial vacuum is created in the needle on withdrawing the plunger to draw the sample into the syringe. Commercial syringes of this type can be made with capacities of 0.5–1.0 μl.

SAMPLE INTRODUCTION

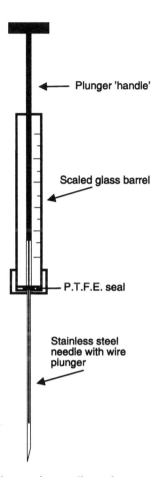

Figure 6.3 Diagram of plunger in needle syringe

There is some controversy with regard to the relative merits of the two types of syringe, although the PIN type syringe is generally preferred for capillary operation. However, high boiling samples are more likely to give discrimination problems with this type of syringe, and so it is worth considering a PIB type syringe if discrimination problems are experienced. Figure 6.4 shows some of the possible errors that can occur on syringe injection into a hot vaporizer.

Grob and Neukom (1979) have studied the influence of the syringe needle on precision and accuracy using vaporization injectors. They describe several types of syringe handling technique with PIB syringes and the discrimination effects associated with them [6]. These techniques are described as **filled needle, cold needle, solvent flush,** and **hot needle methods,** and are identified as follows.

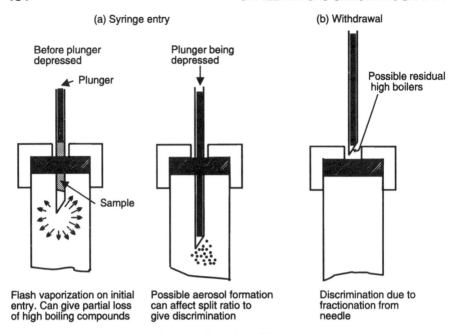

Figure 6.4 Discrimination from syringe handling

1. Filled needle technique (also applies to PIN syringes)
   Sample is drawn into the syringe in the normal manner but the needle remains filled with part of the sample on injection.
2. Cold needle technique
   Sample is drawn into the body of the syringe and the plunger is withdrawn a little further after removing the needle from the sample to fill the needle with air. The sample is then expelled immediately on injection.
3. Solvent flush technique
   The syringe is first filled with pure solvent then the required volume of sample, leaving a small air 'plug' between them. Again the plunger is further withdrawn to fill the needle with air before injection.
4. Hot needle technique
   Sample is drawn into the body of the syringe as before, but now the empty needle is allowed to warm up to the injector temperature for a few seconds before the plunger is depressed.

None of these methods eliminated discrimination altogether for alkanes up to C_{44}, but the hot needle technique gives the least discrimination. The Grob and Neukom study compared these vaporization techniques with that of on-column injection which gave no discernible discrimination. It should be emphasized that discrimination problems arising from incomplete evacuation of the syringe are only likely to arise with very high boiling samples. Provided that appropriate injection techniques and conditions are used, samples should not suffer discrimination effects from either type of syringe. Syringe discrimination effects have also been studied by Schomburg [7–9] who has discussed various features of injector design and technique in relation to minimizing this problem. The other major concern with regard to syringe handling is the transfer of the minute quantity of liquid sample to the GC in an accurate and known amount, although this problem is usually avoided by using the **internal standard procedure** (see Chapter 8).

Summarizing the critical aspects of syringe handling, the following general procedure is recommended.

1. Take the required volume of sample into the syringe.
2. Withdraw the plunger slightly. For PIN syringes this should be by approximately 1 cm to leave an air gap between the bottom of the needle and the sample. For PIB syringes this should be sufficient to ensure that the needle is empty.
3. Quickly wipe the outside of the needle with a clean tissue.
4. Insert the needle into the septum and push it to the full depth. For PIN syringes depress the plunger and withdraw the syringe as quickly as possible. (For splitless and on-column injection see the appropriate sections where a slower injection procedure is recommended.) For PIB syringes use the hot needle technique.

Experience shows that syringe handling should be consistent if a high level of precision is required, and this is best ensured, as mentioned earlier, by using one of the many commercial automatic sampling devices which duplicate the manual procedure except that the handling of the syringe is carried out automatically.

6.3.3 Injection of Gas Samples

Gas samples may consist of permanent gases, volatile liquids which are easily converted to vapours, or permanent gas samples containing compounds at or below their saturation vapour pressures. Samples of the latter type are taken from liquid or solid matrices, either by sampling the headspace or by gas purging, and constitute an important range of applications for capillary GC. These methods can be used to characterize aromas from foodstuffs and beverages, to identify and characterize materials such as geological specimens, to determine trace amounts of contaminants in aqueous samples, and for numerous other purposes. Some examples of on-line instrumentation for these types of analysis are given in Chapter 7.

The simplest way to sample gases is by the use of a gas-tight syringe with a Teflon™ tipped plunger to avoid any leakage of the gas sample from the syringe when it is pressurized in the GC. Sample volumes in the microlitre to millilitre range can be applied to the column in this way using the same type of procedure as for liquid samples, i.e. by injection through a septum. For accurate analysis, however, a calibrated gas sampling valve, known as a **bypass injector**, should be used, filled either from a gas line through which the sample is flowing, or from a gas tight syringe. Several types of commercial sampling valves are available for capillary GC. The most common type uses an external sample loop accurately calibrated to the required volume. For very small volumes (0.1–0.5 μl) fixed internal volume valves are available. In either case, one position of the valve allows the sample volume to be filled which is then transferred into the carrier gas line on operating the valve. All commercial sampling valves can be operated manually, pneumatically or by solenoid, and so this method is ideally suited to process monitoring and automatic operation.

6.3.4 Use of Septa; Septum-less Devices

The **septum** is a critical component of vaporization injectors as it must be thermally stable, reasonably long-lasting, resistant to pressure, leaks and contamination. Since the sample is actually injected by piercing the septum it seems unlikely that all of these requirements can be completely fulfilled. Various forms of silicone polymers are used for GC septa, but there is a wide variety of different types available commercially and users should give careful consideration of their particular requirements before making a selection. Thus, if high temperatures are to be used then appropriate high temperature septa should be selected. Some commercial septa consist of three different layers, the outer layers are harder and have a high temperature resistance, whilst the inner layer is a softer silicone material which ensures effective sealing and longer life. Another variety of septum is faced by Teflon™ surfaces to reduce the possibility of contamination of the system by silicone breakdown products. The latter problem is also minimized by modern injector design which often incorporates a septum purge, previously referred to in Chapter 3. This is a continuous flow of inert gas over the face of the septum and away from the column inlet.

A 'septum-less' device, illustrated in Figure 6.5, is known as the Jade™ valve and is recommended for autosamplers where large numbers of samples are injected repetitiously. Here, the syringe needle passes through the needle guide forming an annular seal. The needle then unseats the ball and enters the injector enabling the sample to be admitted on depressing the plunger. On withdrawing the syringe the ball is returned to its seat by the ball return magnet.

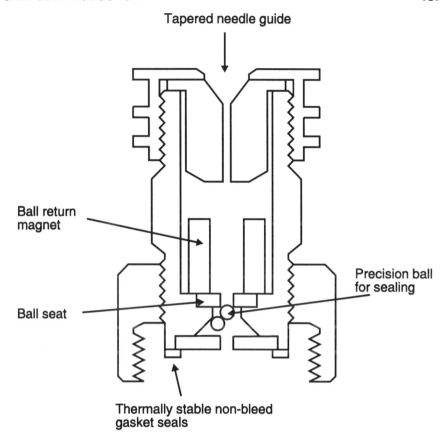

Figure 6.5 Diagram of septum-less injection system: the JADE™ system. (Jade) valve is a trademark of JADE systems, Inc, Austin, Texas, USA; Reproduced by permission of Quadrex Corporation)

6.4 SAMPLE INTRODUCTION METHODS

The requirements listed in Section 6.2 cannot always be realized by using a single method of sample introduction, and so several methods have evolved to deal with particular types of sample. These methods can be listed as follows.

1. Direct injection
2. Split injection
3. Grob-type splitless injection
4. On-column injection (OCI)
5. Programmed temperature vaporization (PTV)

188 CAPILLARY GAS CHROMATOGRAPHY

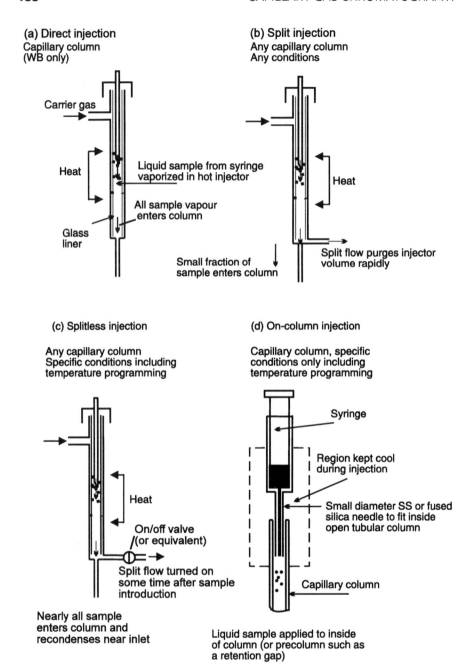

Figure 6.6 Illustration of the injection techniques used in capillary GC

SAMPLE INTRODUCTION

6. On-line thermal desorption and trap (TD)
7. On-line purge and trap injection (PTI)
8. On-line headspace sampling and analysis

Sample introduction into open tubular columns is mostly carried out by techniques 2–4 and these have been studied and described in considerable detail in recent books by Grob [4,5]. Figure 6.6 illustrates the fundamental principles of these techniques.

6.4.1 Direct Injection

IUPAC defines a direct injector as one which directly introduces the sample into the mobile phase stream (Chapter 1, reference 43). Thus the injection of gases or liquid samples without a split flow is considered to be direct injection, and in the case of liquid samples, initial vaporization is a normal feature.

Direct injection is the simplest method of sample introduction and is employed universally with packed columns. For capillary columns, however, the injection band would normally be too long resulting in a considerable loss of column efficiency (Section 6.2.1). Figure 6.7 shows the effect of column diameter on the volume capacities of 25 m columns plotted from (6.8) for two k values.

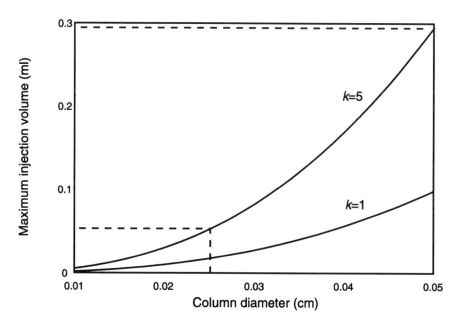

Figure 6.7 Effect of column internal diameter on the volume capacity. Calculated for 25 m × 0.25 mm i.d. columns. Nominal plate number = 80000

Thus for 0.25 mm diameter columns the volume of sample vapour must be less than 0.02 ml for $k = 1$ and less than 0.06 ml for $k = 5$. These vapour volumes are equivalent to liquid volumes of far less than 0.1 μl and would require minute injector volumes, i.e., much less than could be reasonably designed and handled. At a diameter of greater than 0.05 cm, however, the volume capacity of approximately 0.1 ml for even the early components (k > 1) can easily be realized with a 10 cm × 0.1 cm injector liner provided the sample volume is kept to about 0.1 μl. Hence while direct injection is not a feasible option for normal bore open tubular columns, it is possible for wide bore columns. Practical experience, as was shown in Chapter 4, supports this. Direct injection has the following advantages: simplicity; discrimination effects are minimized; and the chances of detecting minor components are maximized because the entire sample is admitted to the column. As indicated in Section 6.2.1, if direct injection is used with programmed temperature operation then the volume capacity can be increased considerably, thereby enabling even larger sample volumes to be injected. Direct

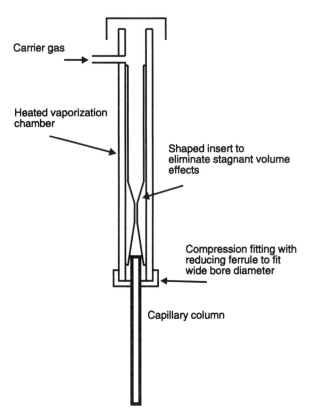

Figure 6.8 Illustration of an effective liner design for direct injection into wide bore columns

SAMPLE INTRODUCTION

injectors of the type normally used with packed columns can usually be adapted for use with wide bore columns simply by making an appropriate choice of glass liner and column fitting. The important features are that the liner should have a compatible volume to that of the volume capacity of the column and that stagnant regions should be avoided. The latter aspect is illustrated by Figure 6.8.

An example of the use of direct injection on a 0.53 mm wide bore column is given in Figure 6.9.

6.4.2 Split Injection

The effective volume of the vaporization injector is easily reduced by purging it with a fast subsidiary flow of carrier gas by means of a bypass situated

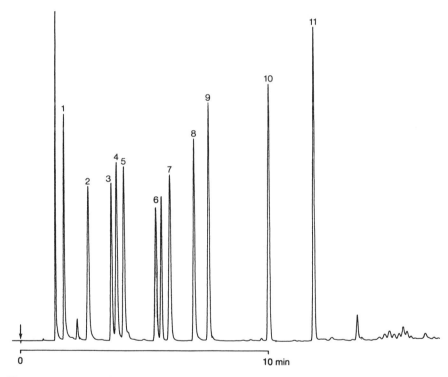

Figure 6.9 Use of direct injection into a wide bore column for the separation of volatile amines. Column: 50 m × 0.53 mm coated with CPsil5CB (5 μm). Temperature 40 (1 min) to 200 °C at 10 °C min−1; hydrogen carrier gas at 165 cm s^{-1}; direct injection. Components: 1 = ethylamine; 2 = *tert*-butylamine; 3 = diethylamine; 4 = *sec*-butylamine; 5 = isobutylamine; 6 = di-isopropylamine; 7 = triethylamine; 8 = 1-pentylamine; 9 = dipropylamine; 10 = di-isobutylamine; 11 = dibutylamine. (Reproduced by permission of Chrompack International BV)

between the injector and the column. This arrangement, now universally known as split injection, is the most commonly used technique for sample introduction in capillary GC.

If the split ratio is S_r then the volume of vapour admitted to the column becomes $v_{inj}/(1 + S_r)$. Also, since the sample admission time is now in the millisecond region, syringe handling time will also play a part. If we assume that the time taken (t_{inj}) to insert the syringe through the septum and depress the plunger is additive to the time taken to purge the injector then the following modification to (6.8) can be derived:

$$v_{inj}(\max) \leq 0.785 d_i^2 \bar{u}(S_r + 1)\left[\frac{2L(1 + k)}{\bar{u}\sqrt{N_c}} - t_{inj}\right] \quad (6.9)$$

A further limitation of split injection is the maximum vapour volume that the injector can accommodate without flooding. However, in contrast to direct injection, the effective injector capacity in split injection is found, in practice, to be larger than its simple geometrical volume because of the simultaneous purging of the injector as the sample is admitted to it. Practical experience shows that the effective capacity of most commercial split injectors is about 1 ml, and so the minimum split ratio required for the column to avoid excessive loss of efficiency is that necessary to purge this volume effectively. Accurate calculations are impracticable in this context as they are affected in a complex manner by the design, geometry, working conditions and the nature of the sample. Nevertheless, a working guide can be obtained by assuming a value of unity for $v_{inj}(\max)$ in (6.9).

Figure 6.10a shows a plot of the minimum split ratio necessary versus column diameter assuming that the maximum tolerable plate loss is 20% and

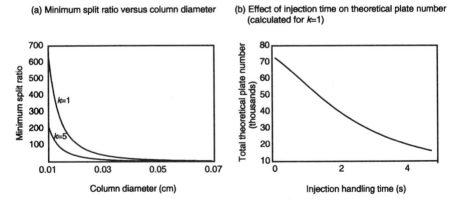

Figure 6.10 Split injection effects. Data calculated for 25 m × 0.25 mm i.d. open tubular column assuming $N = 80000$ theoretical plates

that the injection time is zero. In practice the split ratio should be adjusted to 50–100% higher than that implied from the graph to minimize the plate loss. For instance, 0.25 mm diameter columns should normally be operated with split ratios of about 100:1. Smaller diameter columns require higher split ratios than this, and larger diameter columns can operate with lower split ratios. Wide bore columns do not normally require a split system, as discussed in the previous section. A reasonable balance should be sought between the need to achieve a minimal loss of plate number and to avoid an excessive loss of the sample through the split vent.

Although the syringe handling time (i.e., 'injection time') is assumed to be zero in Figure 6.10a, this can never be so in practice. Whether the injection of the sample is carried out manually or by autosampler it does take a finite time for the syringe to pierce the septum and the sample to be ejected from the needle. Experience shows that a slow injection time using split systems does have a serious effect on the column efficiency and so sample introduction should always be carried out as fast as possible. Equation (6.9) gives an indication of the critical nature of this aspect, as illustrated by Figure 6.10b which shows the loss of theoretical plate number that may be expected from a 25 m × 0.25 mm diameter column for increasing syringe handling time. This depicts the worst possible situation, i.e., for early eluting components ($k = 1$) and, as (6.9) shows, the effect of injection time becomes weaker as k increases. Nevertheless there are many important applications which involve components with low k values and where slow syringe handling will affect the resolution. For instance, a 1 s handling time can reduce the plate number by as much as 30%!

Early forms of split injector had a basic construction which, although satisfactory for qualitative purposes, gave discrimination and sample degradation with labile compounds. It was realized that the geometry of the injector was critical in achieving representative sampling. Samples invariably consist of mixtures of compounds with different polarities and boiling points and so it is impossible to achieve instantaneous volatilization of all the components at the moment of injection. Thus the vapour reaching the split point will have a variable composition as the sample vaporizes and this can cause a change in the split ratio during the critical period of sample introduction, causing in turn a discrimination effect. Discrimination from this source causes a relative loss of the higher boiling components which is a similar effect to that of syringe discrimination. If such discrimination effects are noticed in practice then both possible causes should be investigated.

Injector design is based on the following requirements.

1. Syringe injection through the septum without discrimination.
2. Maintenance of a constant split ratio by use of an appropriate gas control system.

3. Complete and rapid vaporization of the sample and mixing with the carrier gas before it reaches the split point.
4. Provision of a septum purge to avoid septum contaminants from reaching the column.
5. Sufficient temperature range to vaporize all the samples rapidly.
6. Avoidance of any contact with hot metal surfaces which would otherwise cause thermal degradation.

Syringe injection has been discussed in Section 6.3.2 but ways of reducing the possibility of discrimination are included in the design of the injector. One way is to run an injector temperature programme to avoid prevaporization effects (see the 'programmed temperature vaporizer PTV, Section 6.4.4). Another method is to have a cold 'collar' at the top of the injector which surrounds the syringe needle preventing early volatilization, as discussed by Schomburg and Hausig [10].

Modern gas flow systems were discussed in Chapter 3. Figure 3.4 showed a typical system for split injection which involves the use of a mass flow controller to ensure that the volumetric flow rate through the split point is constant. Although this design may help in maintaining constant conditions in general, it is doubtful if it could react at the speed necessary to compensate for the very rapid changes occurring in the injector at the time of sample introduction.

The geometry of the injector is the critical feature since it should create an effective buffer zone to allow sufficient time for complete vaporization and mixing with the carrier gas so that the vapour reaching the split point is homogeneous. Sample degradation problems are avoided by the use of a glass liner which is inserted into the injector and comprises the vaporization chamber.

Figure 6.11 shows a diagram of a commercial split injector.

In this, as in all designs, the carrier gas is preheated in the body of the injector before reaching the liner. The sample is injected into the liner through a septum which is secured by a finned retainer or **septum cap** which keeps it well below the injector temperature. Thorough mixing of the sample vapour with the preheated carrier gas occurs in the porous frit and its arrival at the split vent is delayed by the 'inverted cup' arrangement to minimize discrimination problems. Commercial injectors utilize a variety of liner designs to minimize these discrimination problems involving spiral inserts, glass wool packings and inverted flow configurations. Some commercial liner designs are shown in Figure 6.12. These, and numerous other designs are available in a range of sizes for different instruments. Some of the liners are intended for split operation only, while others, particularly larger volume liners, are for use in splitless and direct injection. In all cases it is important to install the capillary column according to the manufacturer's recommendations.

SAMPLE INTRODUCTION

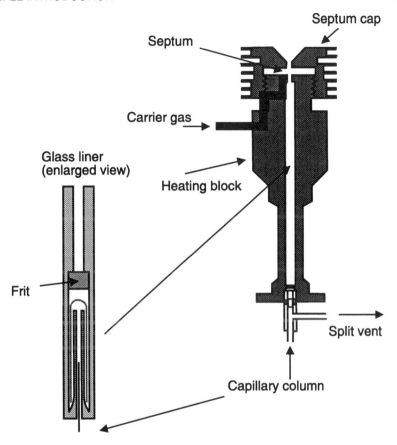

Figure 6.11 Diagram of commercial split injector (Reproduced by permission of Chrompack International BV)

Another important consideration is the condition of the liner. The liner is the sample entry point, operated at relatively high temperatures and so it quickly becomes contaminated with non-volatile matter arising from sample degradation or from material already present. These contaminants inevitably accumulate inside the liner and cause a loss of column performance, usually in the form of peak distortion. It is essential to monitor the liner condition regularly and to either clean or replace it from time to time. A further problem can develop from worn septa due to fragments falling into the liner which can cause a sudden deterioration of peak shape.

6.4.3 Splitless Injection

Split injection is satisfactory for most GC applications and can be employed

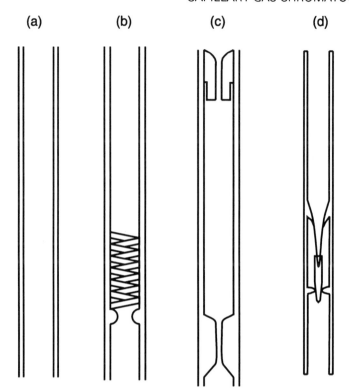

Figure 6.12 Some commercial injector liners used in capillary GC. (a) Basic tubular glass or quartz liner; (b) cyclosplitter™ Restek Corporation; (c) chambered split/splitless liner (R & D Separations); (d) flow inverter™ (R & D Separations)

for any set of column operating conditions. Its main disadvantage is the limitation on sample size which makes the method generally unsuitable for trace analysis. Even under the most favourable circumstances the minimum detectable concentration is in the region of 10 ppm, depending on the type of sample and detector. For instance, organochlorine compounds will often involve the use of the electron capture detector and an appreciably lower detectable limit may be achieved. Concentration levels of less than 10 ppm usually require a different method of sample introduction to enable more sample volume to be introduced to the column.

Splitless injection is a sample introduction technique originally devised by Grob and Grob [11] to enable a much larger volume of the sample to enter the column without the deleterious effect predicted by (6.8). Sample concentrations of less than 1 ppm can then be measured with ease. The splitless technique is generally unnecessary, of course, where direct injection can be applied successfully, as with wide bore columns where all the sample enters the column.

Splitless injection can be carried out manually using a normal split injector fitted with an on/off valve in the split flow line. This arrangement, however, can give rise to tailing peaks due to the long residence time of the sample vapour in the injector which gives it time to diffuse into semistagnant regions which are not normally accessed during the brief residence time associated with split injection. Specially designed systems known as **split–splitless injectors** are available with most instruments and consist of an integrated injector and flow control modules. Split–splitless injectors are designed to avoid these stagnancy effects, and the flow modules have automatic facilities for switching the split flow at the appropriate times. A commercial split–splitless injector is shown diagrammatically in Figure 6.13.

Commercial pneumatic systems can have various designs. The aim is to provide maximum versatility with regard to the use of different injection modes. One example is the automatic system shown in Chapter 3 (Figure 3.5). In the commercial injector shown in Figure 6.13, carrier gas is supplied to the inlet via a pressure regulator and the split vent outlet flow is adjusted by a needle valve and can be turned on and off automatically by means of a solenoid valve. Split vent effluent passes through a low impedance charcoal filter which removes sample components to prevent any contamination of the needle and solenoid valves. The septum purge flow rate is controlled by a fixed impedance frit.

The sample is applied as in split injection, but initially with the split turned off to enable the major fraction of the sample to enter the column. In contrast to split injection, syringe handling during the injection procedure should be

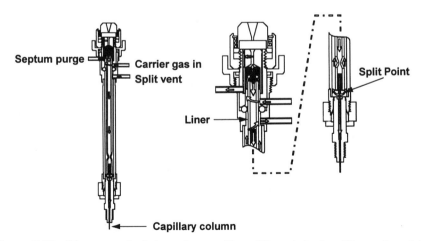

Figure 6.13 Diagram of Autosystem split–splitless injector (Reproduced by permission of Perkin Elmer Corporation)

much slower to reduce the chance of syringe discrimination effects. Column operating conditions are very critical with this method, and temperature programming is essential. The column **must** be set to a sufficiently low temperature initially to ensure that the sample solvent recondenses when it reaches the column inlet region, and the sample components must be less volatile than the chosen solvent, with boiling points at least 20 °C higher. A set period of time is allowed to elapse following sample introduction to allow most of the sample vapour to be transferred to the column. This period, known as the 'splitless time' must be determined by pilot experiments during the development of the method. If this period is too short the sample will be too small, and if it is excessively long the sample will start to migrate along the column to produce a 'smearing' effect with a strongly tailing solvent peak and many sample component peaks will be obscured by the solvent tail. The resulting condensed film of solvent and its entrained components is known as the **flooded zone** and usually occupies 20–30 cm of the column per microlitre of sample injected.

At the end of the splitless time the split flow is turned on to completely eliminate any residual sample vapour from the injector, and at the same time, a temperature programme is commenced. Without this final purge the effect would be even worse than that described earlier for long splitless times. The presence of the condensed film of sample and the rising column temperature causes a solute concentration process to occur which is illustrated by Figure 6.14.

Pure carrier gas entering the column becomes saturated with the solvent at the beginning of the condensed zone, thereby removing solvent and reducing the length of the flooded zone. The carrier gas becomes rapidly saturated with solvent and so the end position of the zone remains almost stationary. The saturated carrier gas travels through the column initiating the start of the solvent peak shortly after the start of the run. As solvent is removed from the front edge of the condensed film so the less volatile sample components redistribute in the residual solvent film, thus focusing the components into a narrow discrete band. Consequently, when all the solvent has been removed the sample components remain as a sharp zone near the start of the column ready to commence their chromatographic migration. The entire process is represented by Figure 6.15 which also illustrates the essential part played by the temperature programming process.

Sample components travel very slowly during the early part of the programme because of their high initial retention factors, and this aids the solute focusing effect. Grob has termed this focusing process the **solvent effect** [4] and it is now widely employed for many important applications, particularly in connection with environmental analysis. Pretorius and coworkers have proposed a theory of the solvent effect [12] although the utilitarian value of such theory is open to question.

SAMPLE INTRODUCTION   **199**

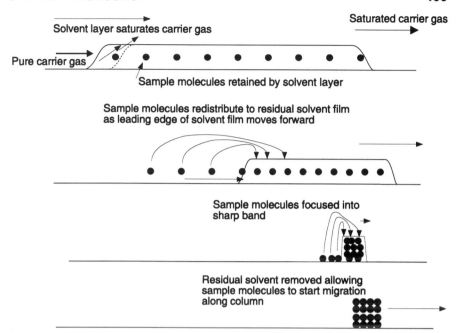

Figure 6.14 Illustration of the solvent focusing effect in splitless injection. [Reproduced by permission of Huethig, based on K. Grob, *On-Column Injection in Capillary Gas Chromatography*, (1986)]

The presence of condensed sample inside the column would be problematical with non-immobilized stationary phases because of their solubility in most organic solvents. To avoid this problem, a length of uncoated capillary tubing known as a **retention gap** should be installed between the injector and the column. The use of a retention gap is recommended even in conjunction with a bonded phase column as it minimizes the chances of peak distortion due to smearing, flooding or film non-uniformity. The length of the retention gap is usually 2–5 m in length depending on the length of the column and the sample size, and its diameter is normally the same as that of the column. Another way of achieving a retention gap is to remove the stationary phase from the first few metres of column by careful solvent rinsing. This, of course, is only possible with non-chemically bonded columns. Although the retention gap does not contain any stationary phase, it must be deactivated to avoid peak tailing effects and it must be wettable by the proposed sample solvent. The latter is accomplished by adapting the deactivation process to produce retention gaps of various polarities, and these are available from commercial suppliers. Coupling the retention gap to the column is carried out using one of the connectors described in Chapter 3, although the use of the glass taper seals is the preferred option.

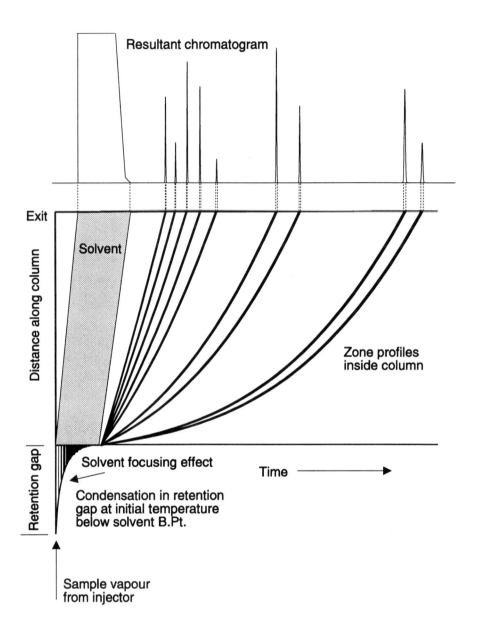

Figure 6.15 The splitless injection process

SAMPLE INTRODUCTION

One of the commonest application of splitless injection is the determination of trace organochlorine pesticides in environmental samples. An example of this is given in Figure 6.16.

The variables involved in splitless injection have been studied by Yang and coworkers using simplex design methods [13] and they have produced some useful working guidelines for the technique. It is suggested that sample sizes of 0.1–10 μl can be analysed successfully by chromatography using a rate of sample addition of about 1 μl s^{-1} and a residence time of the syringe needle (termed 'sampling time') of about 20 s. These sample sizes are appreciably larger than those normally associated with split injection, but it is important to adhere to the author's recommendations concerning liner geometry and syringe handling. Another recommendation is that the splitless time should be greater than 40 s, although this almost certainly relates to very large samples. Modern practice tends to prefer samples rather smaller than 10 μl, preferably 0.1–1 μl, and a splitless time of about 30 s. Also, because of the slower sample application, injector temperatures can be somewhat lower than in split injection, and the authors claim that injector temperatures higher than 230 °C should be rarely necessary. Other recommendations were that the initial column temperature should be 15–30 °C below the boiling point of the solvent, and

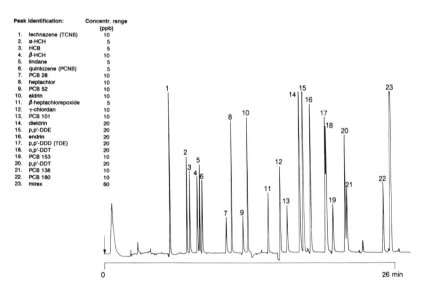

Figure 6.16 Use of splitless injection for the analysis of trace pesticides and PCBs. Column: 50 m × 0.32 mm FSOT coated with CPsil8CB (0.12 μm). Temperature 140 (2.5 min) to 244 °C at 4 °C min^{-1}, then to 290 °C at 20 °C min^{-1}; helium carrier gas; splitless injection; electron capture detection. Sample size = 5 μl. (Reproduced by permission of Chrompack International BV and Ing. W. Brodacz, Bundesanstalt für Agrarbiologie, Linz, Austria)

there should be at least 25 °C difference between the boiling points of the solvent and the lowest boiling component to be determined.

It must be emphasized that these suggestions are working guidelines for splitless injection and the variables should be tested thoroughly in practice to achieve the required level of success.

6.4.4 Programmed Temperature Vaporization Injection (PTV Method)

The problems of syringe discrimination and decomposition in the vaporization techniques described so far obviously could be reduced if much lower sample injection temperatures were employed. One way of avoiding syringe discrimination is to cool the immediate region surrounding the syringe needle by a secondary air flow as described by Schomburg and Hausig in the **cooled needle split** (**CNS**) technique [10]. The programmed temperature vaporizer (PTV), initially described by Poy and coworkers, is a complete injector which can be rapidly heated and cooled allowing the sample to be introduced in liquid form at cool temperatures and then to be vaporized rapidly [14]. Numerous varieties of this type of injector are commercially available, most of which enable both split and splitless modes and in some cases on-column injection, depending on the type of liner chosen. Injector designs for the PTV are usually similar to non-programmable versions except for the inclusion of rapid heating and cooling facilities. These facilities may include the ability to use different programming ramps and cryogenic cooling. PTV injectors are often operated with packed liners depending on the type of application. Thus, the packing may consist of glass or silica wool or adsorbent material to selectively remove certain materials.

While the use of programmable injectors may be advantageous in some applications the merits should be thoroughly assessed by the potential user because in most cases perfectly acceptable results can be achieved with conventional vaporization injectors. Split injection, in particular, is unlikely to be successful using a programmable injector unless temperature programming is applied to the column in order to utilize the focusing effect of the programme and hence to minimize the effect of the vaporization time. Programmable injection, however, is a useful additional facility and may offer advantages particularly in the separation of very high boiling or labile compounds.

Table 6.1 shows a comparison of results using a PTV in two split modes, namely, hot, as in a normal split injector, and then cold for injection followed by programmed 40–300 °C. The sample is a mixture of C10–C28 alkanes covering an atmospheric boiling range of 170 °C to well above 400 °C, i.e. considerably in excess of the highest vaporizer temperatures normally possible in the injector.

Table 6.1 Comparison of PTV hot and programmed modes for discrimination effects. Measured concentrations % w/w

Compound	Theoretical	Hot mode	Programmed mode 40–300 °C Solvent	
			n-hexane	Dichloromethane
C10	12.5	15.2	12.9	12.8
C12	12.5	15.5	12.7	12.7
C14	12.5	15.6	12.7	12.7
C16	12.5	14.6	12.8	12.8
C18	12.5	12.9	12.6	12.6
C22	12.5	10.9	12.5	12.5
C24	12.5	8.7	12.1	12.0
C28	12.5	6.8	11.7	11.7

Reproduced by permission of Chrompack International BV.

Clearly the programmed method gives far less discrimination than the normal hot split injection procedure, almost certainly due to needle discrimination as described earlier. The discrimination rapidly worsens above C18 which boils at 318 °C and if attention is restricted to the range C10–C18 then the hot split injection mode is probably satisfactory.

6.4.5 On-column Injection

On-column injection can be defined as the application of the liquid sample directly to the column, or to a retention gap connected to the column. The technique is often described as **cold on-column injection** to emphasize the necessary liquid state of the sample during its application to the column. Grob has written an extensive treatise on the topic of on-column injection [5], and theoretical treatments have been proposed by Pretorius and coworkers [12], and by Deans [15]. The cited publications indicate that there is some considerable controversy surrounding the mechanism of the technique, and potential users could easily be put off by its apparent complexity. However, the method can be used very successfully with modern commercial injectors provided certain ground rules are followed. In fact, on-column injection is probably the best method of sample introduction for capillary columns from the viewpoints of quantitation and trace analysis. If the injection process is carried out effectively then the errors associated with flash vaporization techniques are eliminated. Also, very high boiling compounds, which can give problems due to incomplete vaporization with vaporization techniques, are easily admitted by on-column methods. Unfortunately, because of the time scale necessary to vaporize the sample and eliminate the solvent, on-column injection is only normally feasible when used in conjunction with programmed temperature operation. Thus the method in many ways is similar to splitless injection but without the vaporization and recondensation stages.

Various practical techniques have been applied to achieve this purpose, mostly based on using a syringe needle of sufficiently small diameter to enter the bore of the column. The stainless steel needles used for piercing septa normally have an external diameter of about 0.5 mm, which is too wide to enter the majority of open tubular columns.

In the case of wide bore columns, on-column injection is possible with stainless steel needles but preferably with a slightly smaller diameter than 0.5 mm. On-column injection into wide bore columns is sometimes achieved using vaporization injectors fitted with a special liner which allows the column inlet to protrude sufficiently into the injector so that the syringe needle enters it on its insertion through the septum. This is facilitated by means an hourglass constriction for the liner similar to that used for glass taper type connections. The injector can be operated under a variety of temperature conditions ranging from 'hot' to 'cold' on-column, although in the latter event, the injector must be provided with a programming facility.

It is, of course, possible to carry out on-column injection onto smaller bore columns using a similar technique but incorporating a short length of wide bore precolumn between the actual column and the injector to accept the sample entry. This principle has in fact been employed in several commercial devices, including autosamplers.

The more usual form of on-column injector involves the use of a syringe fitted with a small diameter needle which can enter the column bore. Also, because of its small size the needle must be able to reach the column inlet without undue stress such as would be applied when penetrating a septum. Lengths of flexible fused silica tubing are normally used as on-column syringe needles, and these are inserted into the column by means of an injector which accurately aligns the needle with the column bore. Various methods are employed to allow access of the syringe needle to the column based either on some form of pneumatic or electrically operated mini-valve, or a plastic seal or O-ring which is normally opened manually as part of the sample injection procedure. On-column injectors are also equipped with means of cooling the device during the injection period, usually by a secondary air flow. This is essential to ensure that no prevaporization of the sample can occur. Undoubtedly, correct installation of the column is more critical with on-column injection than with other techniques, and the injection procedure requires rather more care. A diagram of a commercial on-column injector is shown in Figure 6.17.

This injector is designed to enable both manual and automatic sample introduction into column diameters in the range 100–530 μm, and can be temperature programmed independently of the column conditions. In this case the syringe needle penetrates a self-sealing septum but the geometry is designed to give efficient sample transfer to the column.

SAMPLE INTRODUCTION 205

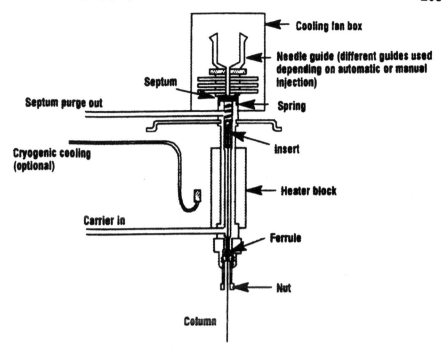

Figure 6.17 Diagram of commercial on-column injector, The Hewlett-Packard 5890 series 11 on-column injection port. (Reproduced by permission of Hewlett-Packard Ltd)

Figure 6.18 shows the use of this injector for the analysis of Triton X-100, a non-ionic detergent.

Critical features of on-column injection

Although on-column injection can be very successful in many cases, it does demand a higher level of understanding than split injection, with more manipulative skill. The commonest problems are peak distortion effects arising mainly from so-called **band spreading in space**. Band broadening in space describes a range of general effects which are described as follows.

(a) The sample is ejected from the syringe too fast giving minute droplets which produce a heterogeneous and extended flooded zone. This becomes more of a problem with large sample sizes (namely, greater than one microlitre) and gives distorted peaks, which often show shoulders or several maxima, known as **split peaks**. A slower injection action should be employed if necessary. Figure 6.19 shows an example of this effect.

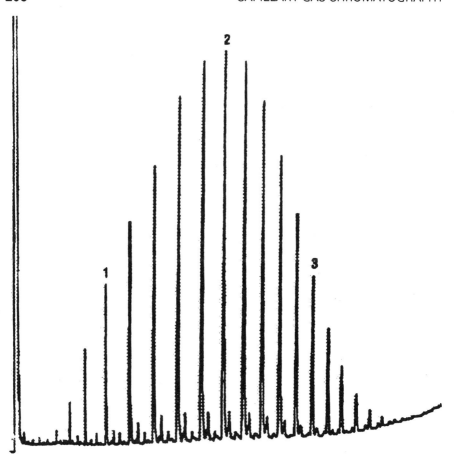

Figure 6.18 On-column injection of Triton-100. Column: 10 m × 0.53 mm aluminium clad coated with 0.1 μm HP-1 (A high temperature polysiloxane phase). Pressure programmed 4–7 psi at 0.1 psi min^{-1}. Injection volume 1 μl cool on-column, programmed as column oven (oven track mode) (Reproduced by permission of Hewlett-Packard Ltd)

(b) Sample solvent does not wet the column inlet or retention gap, causing it to break up into droplets and migrate along the column. This could occur, for instance, if the column is coated with a different polarity phase to that of the sample solvent. Remember that the phase must be immobilized if a retention gap is not used. If a retention gap is used then this should be deactivated to give a compatible polarity to that of the solvent, as described for splitless injection.

(c) Gravity effects in the separate coils of the column (i.e., the solvent drains to the bottom of each loop to give several sub-zones). This is only likely to arise in the case of very large samples.

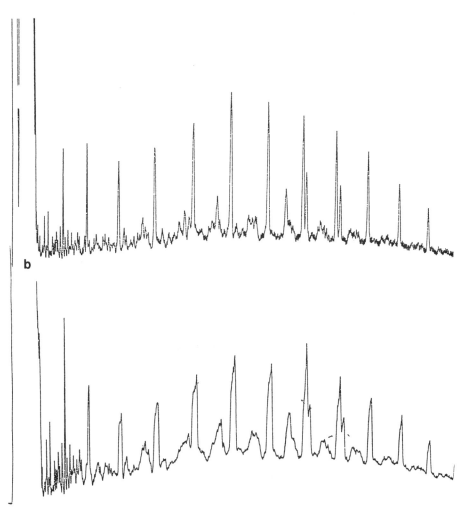

Figure 6.19 Peak distortion due to band spreading in space as a result of the injection speed in on-column injection. (a) Slow injection and (b) fast injection

(d) Premature evaporation from the front (i.e., the column side) of the flooded zone. As we discussed earlier, a successful solvent effect depends on evaporation from the back of the flooded zone by saturation of the carrier gas. Premature evaporation from the solvent front would be caused by the initial column temperature being too high. Nevertheless, an initial column temperature of slightly above the solvent boiling point is often recommended; this is slightly higher than that recommended for splitless injection. The reason for this difference is that in on-column injection the sample is applied to the column very much more quickly and the use of a higher initial temperature causes **concurrent solvent evaporation (CSE)** which creates a back pressure on the column side of the flooded zone thereby preventing an excessive spread of the liquid solvent into the column. This limits the length of the flooded zone and aids the solvent focusing effect. In fact, this process is utilized as a way of introducing large samples into a capillary column, as in combined LC–GC. (see Section 7.5.2). In practice in on-column injection, there is probably an element of both the solvent effect as described for splitless injection and CSE, and it is necessary to carry out some preliminary trials with new applications to find the best conditions.

Most of the problems associated with this method of sample introduction can be avoided by the use of a retention gap, as for splitless injection. This should again have the same diameter as the column, and be deactivated with a compatible polarity. The injector itself should be cooled sufficiently to avoid any possibility of solvent evaporating at the point of its introduction so that vapour diffuses back into the inlet to cause severe peak tailing of the solvent and component peaks.

As mentioned earlier, this secondary cooling is usually accomplished by an air flow around the injector body. In some versions of on-column injector the device is moveable so that it can be physically raised above the column oven; this allows the sample to be injected at ambient temperatures. The injector is then lowered into the oven for the injection procedure to continue.

Another important feature of on-column injection is the method of gas flow control. This should be designed to counter the carrier gas back-flow which occurs when the injector valve, or seal is opened. If the on-column injector does not have a subsidiary gas flow then there is a considerable increase in flow rate when the injector valve, or seal is opened. Conventional pressure regulators cannot quickly adjust the inlet pressure under these circumstances and so there would be a considerable time lag before the pressure is restored. One method of obviating this problem is to incorporate a purge in the injector so that the pressure regulator is supplying gas at a reasonable flow rate continuously. When the injection valve is opened then the increase in flow rate is proportionally reduced and allows the regulator to readjust much more quickly.

On column injection enables samples as large as several microlitres to be introduced to the column, and so trace components at the sub ppm level can be detected and measured. Hinshaw and Yang have reported that sample sizes in excess of eight microlitres can be applied with an on-column injector which is temperature programmed independently from the column oven [16].

Greater care is needed with this technique to avoid column contamination from any solid or involatile material as this will remain at the point of injection causing an eventual deterioration of the column. This is another good reason to use a retention gap as it is less expensive to replace this than the column. Nevertheless, an effective sample preparation procedure should always be applied wherever possible to remove contaminants and other undesirable materials.

6.5 GENERAL SUMMARY

Split injection is the standard method of sample introduction in capillary GC. The technique can be used with almost any type of sample under any column conditions, and usually gives satisfactory results except in trace analysis and the analysis of very high boiling compounds. Modern injector and flow control design have minimized many of the quantitative problems previously associated with this technique, although operator technique and chromatographic conditions are still critical features. Split injection is certainly the simplest method of sample introduction from the operator's viewpoint, and is the method most widely used by autosamplers.

Direct injection is used commonly with wide bore columns, but is generally unsuitable for normal or narrow bore columns. Where direct injection can be used it provides several important advantages, namely, all the sample is introduced to the column enabling trace analysis, and this also avoids the need to use dedicated capillary facilities.

The Grob method of splitless injection is used exclusively for trace analysis as a way of introducing much larger sample volumes to the capillary column without producing a loss in column efficiency. The method is widely employed in environmental analysis, but it does involve a number of variables and so considerable expertise is necessary in setting up the analytical conditions. The procedure should be straightforward, however, for users of standard methods in which the variables have already been standardized.

On column injection is the best recognized method of sample introduction for the capillary column, because it avoids the excessively high injector temperatures associated with vaporization techniques. This removes the primary cause of discrimination in quantitative analysis and reduces the likelihood of sample decomposition. The main disadvantage is that programmed temperature operation is usually essential, and so the method is unlikely to be useful for very volatile samples.

REFERENCES

1. Lee, M., Yang, F. J. and Bartle, K. D. (1984) *Open Tubular Column Gas Chromatography: Theory and Practice*, Wiley, New York.

2. Jennings, W. (1987) *Analytical Gas Chromatography*, Academic Press, Orlando.
3. Sandra P., (ed) (1985) *Sample Introduction in Capillary Gas Chromatography*, Vol. 1, Heuthig, Heidelburg.
4. Grob, K. (1986) *Classical Split and Splitless Injection in Capillary Gas Chromatography*, Huethig, Heidelburg.
5. Grob, K. (1987) *On-Column Injection in Capillary Gas Chromatography*, Heuthig, Heidelburg.
6. Grob, K. and Neukom, H. P. (1979) *J. High Res. Chromatogr. Chromatogr. Commun.*, **2**, 15.
7. Schomburg, G. (1977) *Gaschromatographie*, Verlag Chemie, Weinheim.
8. Schomburg, G. Behlau, H. and Dielmann, R. (1977) *J. Chromatogr.*, **142**, 87.
9. Schomburg, G. (1987) *Separation*, March, 3.
10. Schomburg, G. and Hausig, U. (1985) *HRC&CC*, **9**, 572.
11. Grob, K. and Grob, G. (1969) *J. Chromatogr. Sci.*, **7**, 584.
12. Pretorius, V., Phillips, C. S. G. and Bertsch, W. (1983) *HRC&CC*, **6**, 273.
13. Yang, F. J, Brown, A. C. and Cram, S.P. (1978) *J. Chromatogr.*, **158**, 91.
14. Poy, F., Visani, S. and Terrosi, F. (1981) *J. Chromatogr.*, **217**, 81.
15. Deans, D. R. (1971) *Anal. Chem.*, **43**, 2026.
16. Hinshaw, Jr, J. V. and Yang, F. J. (1983) *HRC&CC*, **6**, 554.

7
Sample Preparation

7.1 INTRODUCTION

Capillary GC is widely used for the analysis of samples derived from a diverse range of sources, many of which are too complex to be sampled directly. In these situations the volatile components must be extracted to produce a representative sample which may then need further treatment before it is suitable for application to a capillary column. In fact, the preparation procedure could involve several stages of extraction, cleanup, concentration and perhaps further selective extraction.

Another requirement is that the sample should be representative of the original material, and a statistical sampling scheme may need to be applied, as would be the case for instance in environmental monitoring. This then becomes more of a sampling problem rather than one of sample preparation and is outside the scope of the present book. Many of these techniques and methods of sample preparation have been described by Grob and Kaiser [1].

Table 7.1 is a summary of procedures that can be employed for sample pretreatment, and the types of sample associated with these methods.

Sample preparation can be the most time consuming part of the analysis, involving perhaps several of the procedures listed in Table 7.1, particularly for samples of biological or environmental origin. However, there is now an increasing trend towards the development and use of small scale methods such as micro-distillation, solid phase extraction, and even of on-line techniques in which extraction, concentration and analysis are performed automatically by a single integrated instrument.

7.2 DISTILLATION PROCEDURES

Many samples are amenable to distillation procedures, and this should certainly be carried out wherever possible. Thus, restricting the boiling range

Table 7.1 Sample preparation techniques

Technique	Typical sample type	References
Fractional distillation	Wide boiling range liquids, petroleum oils, hydrocarbon oils	2, 3
Vacuum distillation	Residues and involatile matrices	
Steam distillation (Dean and Stark, etc.) micro-steam distillation	Liquids and solids to isolate volatile constituents	4
Sublimation	Polycyclic hydrocarbons	5
Freeze drying	Biological, food samples, thermally labile compounds	6
Solvent extraction, (Kuderna–Danish concentrator Soxhlet)	Gases, vapours, headspace gases, aqueous samples, organic liquids	7, 8
Ion pair extraction	Aqueous samples	
Supercritical fluid extraction (SFE)	Solids (soils, particulates) acids and bases	9, 10
Preparative scale chromatography	Fractionation according to polarity, solubility, etc., to recover discrete fractions for further separation	See text
Solid phase extraction (SPE) using adsorption or reverse phase bonded columns	As above	11, 12
Size exclusion on macroreticular porous polymers	Molecular weight fractionation	13
Liquid chromatography coupled with GC	General fractionation technique, not yet widely employed	14, 15
Multidimensional GC	Heart-cutting fractionation complex analyses	16
Headspace analysis static, dynamic, and purge and trap techniques	Environmental samples, water, beverages, foodstuffs, soils, geological specimens, polymers, solids, etc.	1, 16–19

SAMPLE PREPARATION

of the sample will prolong the column lifetime and produce more accurate quantitative results. Depending on the type and complexity of the sample, distillation may be applied to retrieve the total volatiles leaving an involatile residue, or to produce narrow boiling range fractions from wide boiling range samples.

Steam distillation is a well-known technique, widely employed as a method for the recovery of compounds immiscible in water. The technique is based on the fact that the total vapour pressure of two immiscible liquids is the sum of the individual vapour pressures. Thus, many compounds having much higher boiling points than water can be entrained in steam when distilled in the presence of water using relatively simple apparatus. The **Dean and Stark** apparatus is well known for laboratory scale operations but micro-scale apparatus which is more compatible with capillary GC is available from commercial suppliers.

7.3 SOLVENT EXTRACTION

Solvent extraction is used widely for recovering volatiles from gases, liquids and solids. Volatiles in atmospheric samples are usually isolated by drawing a known volume of air through a bottle via a sinter or bubbler immersed in an appropriate solvent. A flow meter is incorporated in the system to measure the total volume of air. Provided that the volatile compounds are soluble in the solvent their subsequent measurement enables the original concentrations in the atmosphere to be determined. However, this method has been largely replaced by the use of solid absorbents from which the volatiles can be recovered by thermal desorption techniques (see Section 7.5).

Solvent extraction is used more commonly as a way of isolating volatiles from aqueous and solid samples. Techniques range from simple manual extractions by separating funnels to the use of systems which carry out dynamic continuous extractions with a high degree of efficiency, such as the **Kuderna–Danish** evaporative concentrator. Whatever method is employed, it is essential to choose a solvent which favours the components to be extracted, and in the case of manual extractions, to repeat the process several times. This is because the concentrations of each extracted solute in successive extraction stages follow a binomial expansion characteristic. Thus if, for example, an initial extraction recovers 80% of any solute, 96% would be recovered on two extractions, and 99.2% for three extractions.

The common laboratory procedure of **Soxhlet** extraction is a popular and effective method for extracting volatiles from solid samples but is time consuming and could be regarded as somewhat dated in view of the modern thermal desorption and supercritical fluid extraction techniques that are now available.

There is a considerable awareness nowadays with regard to the health and safety aspects of solvent use and this has led to the development of methods which utilize only small quantities of solvent, or alternative techniques which avoid their use altogether such as on-line purge and trap and headspace methods.

The analysis of aerosols such as tobacco smoke, stack gases, atmospheric particulates, etc., is another important application area for capillary GC, and sampling of these situations presents a difficult challenge to the analyst. Some of these problems, and the techniques employed to overcome them have been described by Grob and Kaiser. A favoured technique is to use millipore filters to collect the particulate matter, and the soluble organic material is then solvent extracted [1].

7.4 SUPERCRITICAL FLUID EXTRACTION (SFE)

Gases above their critical pressure and temperature are in a supercritical state, intermediate between that of a true gas and liquid. Supercritical fluids have strong extraction properties because the 'solubility' of compounds in the fluid is close to that of a true solvent, and the much lower viscosity allows it to percolate easily through packed beds of sample. Thus, not only is there an efficient contact between the extracting fluid and the sample but the fluid is easily removed when it is released from its supercritical state.

Carbon dioxide is nearly always the chosen gas for SFE in view of its innocuous nature and mild critical conditions, namely, a critical pressure of 75 bar and a critical temperature of 31 °C, values which are relatively easy to achieve in modern equipment.

There are several commercial systems that combine SFE with a capillary GC. Figure 7.1 shows a diagrammatic view of a system which includes a facility for the extraction and analysis of multiple samples.

The sample holder contains a number of small stainless steel cartridges which are filled with the samples in a particulate state. Solid samples such as soil or sediments are packed into the cartridges without any pretreatment, and aqueous samples can be flushed through the cartridges filled with an appropriate absorbent to concentrate all of the contaminants. The cartridges are sequentially fed into the extraction oven and the carrier gas line which at this stage consists of supercritical carbon dioxide. The extracted compounds are carried to the cold trap and condensed after the heated constriction which restores the CO_2 to its true gaseous state. After an appropriate extraction period the circulation of the coolant ceases and the trap is rapidly heated to vaporize the components. At the same time the column temperature is programmed according to the required conditions. The adjustable split, partitions the sample size to avoid the possibility of severe overloading effects.

SAMPLE PREPARATION

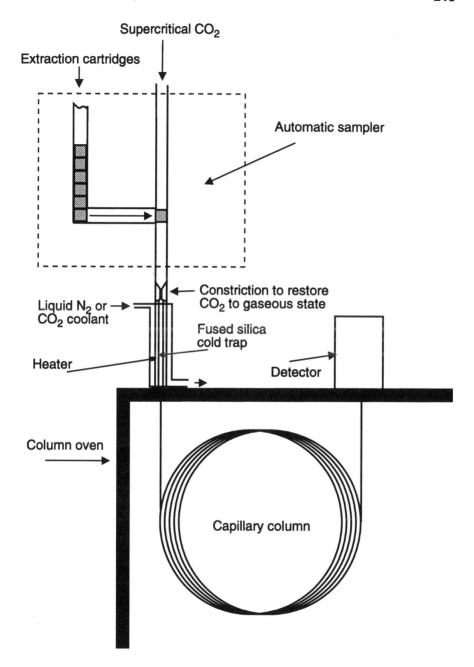

Figure 7.1 Schematic diagram of commercial SFE–GC system. (Reproduced by permission of Chrompack International BV)

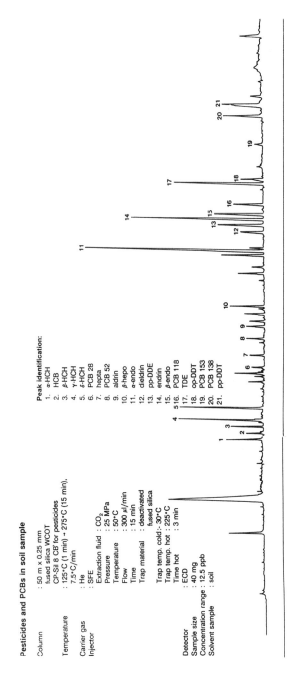

Figure 7.2 Example of SFE–GC combination for environmental analysis

Figure 7.2 shows the application of this apparatus and the relevant conditions for the analysis of pesticides and PCBs in a soil sample.

7.5 EXTRACTION BY SOLIDS

Solid absorbents are widely used in sample preparation for the analysis of gases, vapours and liquids, and particularly aqueous samples. The techniques include component type separation using solid phase extraction and the total absorption of volatiles using carbon, polymers, etc. Under this heading we can also include the many varieties of chromatography namely, adsorption, HPLC and size exclusion methods. These methods are now preferred to bulk liquid extraction methods for the reasons stated earlier, and because they are easily adapted to on-line automatic procedures.

Preparative scale adsorption chromatography is a classical method of sample cleanup and prior fractionation according to sample type. Silica gel and alumina are the most commonly used adsorbents and a typical application is for polycyclic compounds in complex oils and fuels [20]. Carcinogenic polycyclic aromatic hydrocarbons (PAH) are often associated with environmental situations such as emissions, smoke, etc., and an efficient preparation procedure is essential, often involving preliminary adsorption chromatography [21]. The technique of **size exclusion chromatography** can also be used to advantage as a prepreparation technique particularly for the prior fractionation of oils, fats, environmental samples, etc., into discrete molecular weight fractions, and as a way of removing very high molecular weight material from complex samples.

7.5.1 Solid Phase Extraction (SPE)

Solid phase extraction is rapidly becoming a popular method of sample pretreatment for all forms of high resolution chromatography. The method is based on the use of small tubes, or cartridges filled with 100–500 mg of an appropriate absorbent chosen from a wide range of available materials. There is, however, a recently developed type of SPE cartridge in which very small particles of the absorbent are enmeshed in a web of PTFE microfibrils. These have a similar collection efficiency to the packed tubes but require less pressure drop [22,23].

The sample is applied to the SPE cartridge either from a syringe, or a sampling manifold, and forced through the absorbent bed either by applied pressure or suction at its lower end. Components are retained, usually according to their type, and are concentrated inside the cartridge. They are subsequently recovered by solvent flushing. The method is attractive because of the small amounts of sample and materials necessary, and the much greater speed of the procedure compared with classical adsorption methods.

Table 7.2 Characteristics of silica bonded phase sorbents

Sorbent type	Sample type	Typical applications
Octadecyl	Reverse phase extraction of non-polar compounds	Drugs, essential oils, food preservatives, vitamins, plasticizers, pesticides, steroids, hydrocarbons
Octyl	Reverse phase extraction of moderately polar compounds. Compounds bound too tightly to octadecyl silica	Priority pollutants, pesticides
Phenyl	Reverse phase extraction of non-polar compounds. Provides less retention of hydrophobic compounds	Does not seem to be widely used
Cyanopropyl	Normal phase extraction of polar compounds	Amines, alcohols, dyes, vitamins, phenols
Silica gel	Adsorption of polar compounds	Drugs, alkaloids, mycotoxins, Amino acids, flavinoids, heterocyclic compounds, lipids, steroids, organic acids, terpenes, vitamins
Diol functionality	Normal phase extraction of polar compounds (similar to silica gel)	proteins, peptides, surfactants
Aminopropyl	Normal phase extraction	carbohydrates, peptides, nucleotides, steroids, vitamins
Dimethylaminopropyl	Weak anion exchange extraction	amino acids

Reproduced by permission of Elsevier Science, from C. F. and S. K. Poole (1991) *Chromatography Today*, p. 779.

A summary of some of the available sorbents for SPE cartridges and their typical applications is given in Table 7.2.

7.6 HPLC–CGC COMBINATION

For many years the combination of HPLC with capillary GC was considered to be impracticable because of the incompatibility of the phases and the relatively large volumes of solvent associated with HPLC fractions. Nevertheless, direct coupling of the two techniques offers a potentially powerful method of sample treatment. This would allow complex samples to be prefractionated rapidly according to their molecular weight or chemical classification on a suitable HPLC precolumn, and appropriate fractions then separated by on-line capillary GC.

The problem of coupling the outlet from an HPLC instrument directly to the inlet of a capillary gas chromatograph is accomplished by a bypass valve fitted

Table 7.3 Typical applications of LC–GC

Sample analysed	Analytes
Sorghum	Atrazine
Butter	Pesticides
Coal tars	PACs, PCBs
Petroleum fuels	PACs, chemical classes
Urine	Diethylstilbestrol
	Heroin
	Metabolites
Toothpaste	Dyestuff
Raspberry sauce	Raspberry ketone
Olive oil	Wax esters
Hops	Pesticides
Aqueous samples	PCBs, pesticides
Fish extracts	PCBs
Diesel particulate extracts	PACs
Plasma	Broxaterol
Triglycerides	Plasticizers
Shale oil, lignite tars	Chemical classes
Bacteria	Triglyceride esterification products

Reprinted with permission from I. L. Davies, M. W. Rayner, J. P. Kithinji, and K. D. Bartle (1988) *Anal. Chem.*, **60**, 683A. Copyright (1988) American Chemical Society.

with a sample loop of appropriate volume. Selected fractions are passed to the capillary column via a large retention gap (see Chapter 6) which utilizes the process of concurrent solvent evaporation to retain the fraction in the retention gap thus avoiding any preliminary contact with the column. This process, described earlier in Chapter 6 in connection with on-column injection, balances the rate of solvent addition to the retention gap with the rate of evaporation from the solvent front. Thus the evaporation of the solvent produces a back pressure on the column side of the sample thus restricting the length of the flooded zone. The column temperature must, of course, be adjusted to slightly above the boiling point of the HPLC solvent which must consist only of volatile components.

A useful description of the methods used for interfacing both SFE and HPLC to capillary GC has been published by Bartle and his coworkers [9]. Table 7.3 gives some of the typical possible applications of this combined technique and Figure 7.3 shows a typical application.

7.7 HEADSPACE ANALYSIS

Headspace analysis can be loosely defined as the determination of characteristic volatile compounds associated with liquids or solids without **direct**

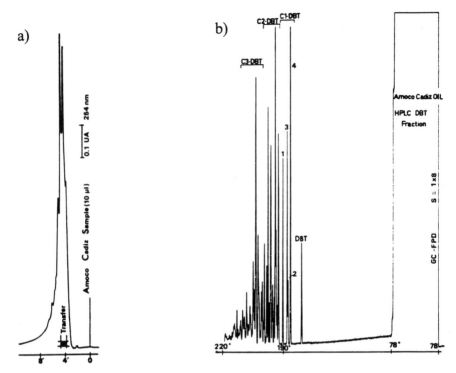

Figure 7.3 Use of HPLC–GC combination for determination of dibenzothiophenes (DBT) in crude oil. (a) HPLC chromatogram showing DBT fraction and (b) GC chromatogram of DBT fraction. [Reproduced by permission of Huethig, from F. Berthou and Y. Dreano, *HRC&CC*, **11**, 706 (1988)]

sampling of the matrix. It is regarded, therefore, as a secondary method from which the original component concentrations can be determined by calibration. Typical samples can include water, effluents, soil, food and beverages. The purpose of the analysis can be to evaluate and identify individual components or to relate chromatographic profiles to odour or aroma characteristics. Not surprisingly, most manufacturers provide automatic equipment for many of these applications, and these include air analysers, water analysers, automatic headspace analysers, thermal desorbers, and purge and trap analysers.

The sample itself is often in an unsuitable state for direct GC and so without headspace sampling it would require considerable sample pretreatment and cleanup before it could be applied to a GC column. The volatiles comprising the headspace are therefore sampled and analysed by chromatography, and calibration procedures are employed to convert the measured component concentrations into values associated with the original sample. Several techniques are used for headspace sampling and for the subsequent handling before the analysis.

SAMPLE PREPARATION

7.7.1 Types of Headspace Sampling

There are three basic methods of headspace sampling, namely, **static headspace**, **dynamic headspace** and **purge** methods. These are illustrated by Figure 7.4, which also includes illustrations of the methods of sample recovery: solvent extraction, thermal desorption and cold trapping.

(a) Static headspace sampling

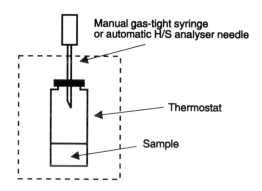

(b) Dynamic headspace and trapping

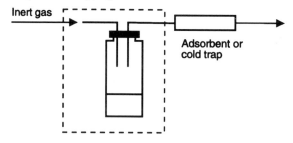

(c) Purge and trapping

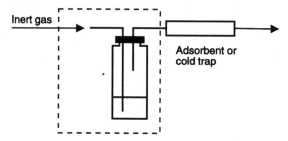

Figure 7.4 Illustration of headspace sampling techniques

7.7.2 Static Headspace Sampling

A basic method of static headspace sampling can be accomplished with a gas-tight syringe by taking a known volume of the headspace gas above the sample. The sample must be thermostatically controlled and allowed to reach an equilibrium situation. The headspace sample should be small in comparison with the total headspace volume to prevent any change in the characteristics of subsequent samples. Gas samples of 1–2 ml can be taken in this way and injected for capillary column analysis provided split injection is used. Alternatively, direct injection could be employed but with a liquid nitrogen cold trap installed at the beginning of the column, similar to the SFE equipment shown in Figure 7.2. The condensed volatiles can then be subsequently volatilized and applied to the column operated under programmed conditions. This method is unlikely to be successful for determining concentrations lower than the parts per million range, because of the relatively small sample volumes associated with static sampling.

Commercial automatic headspace samplers are manufactured by many of the suppliers and these are largely based on the static headspace principle. Samples are usually contained in thermostatically controlled vials in a carousel for multiple analysis, and headspace samples are taken automatically by penetration of a septum by a needle. The Varian system shown in Figure 7.5 is typical of the principles employed. The stages are as follows.

1. Standby: the sample equilibrates in its vial before the sample needle penetrates the septum. Solenoid valve SV1 is open and SV2 is closed to allow the loop and needle to be continuously flushed.
2. Vial pressurization: the needle penetrates the septum pressurizing the sample to the required pressure by the pressurization gas.
3. Loop fill (Figure 7.5a): SV1 closes and SV2 opens. This allows the headspace gas to vent through the sample loop thereby filling it.
4. Injection to GC (Figure 7.5b): the headspace sample is connected into the carrier gas flow line and hence to the column.

The concentration of each component in the headspace is related its concentration in the liquid sample according to **Henry's law** by

$$C_0(\text{liquid}) = [C(\text{gas})(KV_L + V_G)]$$

where C_O = original concentration of component in liquid; $C(\text{gas})$ = concentration in the headspace gas; K = partition coefficient; V_L = volume of liquid; and V_G = volume of headspace gas.

This relationship will only hold provided Henry's law is applicable. For the sake of accuracy, a careful calibration should be carried out over the range of relevant concentrations. A further complication could arise if some of the required compounds interact with the sample matrix to give higher partition coefficients than otherwise.

SAMPLE PREPARATION

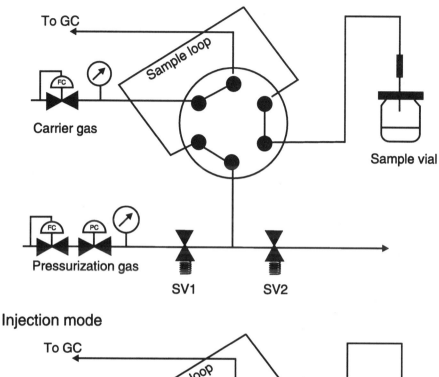

Figure 7.5 Gas flow system of commercial static headspace sampler. (Reproduced by permission of Varian Analytical Instruments)

Figure 7.6 is an example of the use of the Varian headspace sample to measure residual solvents in pharmaceuticals.

7.7.3 Dynamic Headspace Sampling and Purge Techniques

Dynamic headspace sampling involves purging the headspace with a large known volume of inert gas which ultimately removes most of the volatile compounds. Exceptions would be those compounds showing a strong affinity for the sample matrix, such as polar compounds in aqueous samples. The removal of volatiles in this way is described by the following relationship:

$$FR_t = 1 - e^{-(Ft(KV_L + V_G))}$$

where FR_t is the ratio of the amount of any particular volatile removed after time t to its original amount in the sample and F is the purge gas volumetric flow rate. With aqueous systems, the stripping efficiency depends critically on the value of K. Non-polar or medium polar compounds, for instance, with medium or low partition coefficients will be purged quickly, but more polar compounds such as alcohols and carboxylic acids favour the aqueous phase strongly and so they cannot be removed effectively by this method.

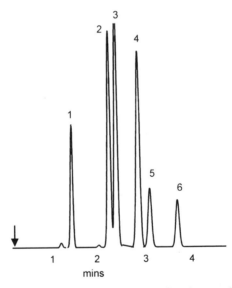

Figure 7.6 Static headspace chromatogram of solvents found as impurities in pharmaceuticals (100 ppm of each component in water). Headspace sample volume: 50 μl at 80 °C. Components: 1 = ethylene oxide; 2 = methylene chloride; 3 = benzene; 4 = trichloroethylene; 5 = chloroform; 6 = dioxane. (Reproduced by permission of Varian Analytical Instruments)

SAMPLE PREPARATION

The exit gas from the sample is 'scrubbed' to recover the volatile compounds either by passing it through a suitable absorbent material, or by condensing the volatiles in a cold trap. Tenax™, a polymer based on 2,6-diphenyl-*p*-phenylene oxide is commonly used for this purpose and is available in several forms. Tenax™ TA, a highly purified form of the polymer is stable to 375 °C and gives insignificant bleed of organics. A material suitable for the recovery of lower molecular weight compounds is Tenax™ GR which contains 23% graphite. This adsorbent is suitable for the efficient trapping of compounds with low to medium polarity and which can then be recovered quantitatively by either solvent extraction or thermal desorption.

Purge techniques are similar to dynamic headspace sampling except that the gas is passed **through** the sample. Clearly, any apparatus used for dynamic headspace sampling could also be employed for purge sampling by using an appropriate sampling vessel.

Both dynamic headspace and purge methods are available in a variety of automatic commercial systems, enabling concentrations in the ppb–ppt range to be measured. Figure 7.7 shows the principle of a commercial system which can be operated either for on-line thermal desorption, or purge and trap analysis. This versatility is enabled by a system of automatic solenoid valves which select the appropriate gas flow paths.

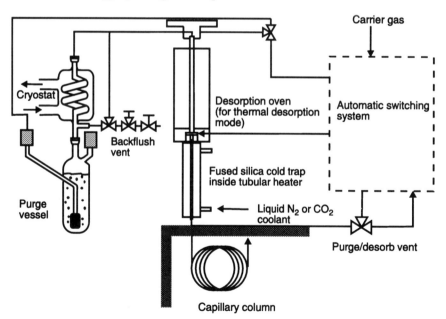

Figure 7.7 Schematic diagram of commercial system for dynamic headspace/purge and trap sampling and analysis. (Reproduced by permission of Chrompack International BV)

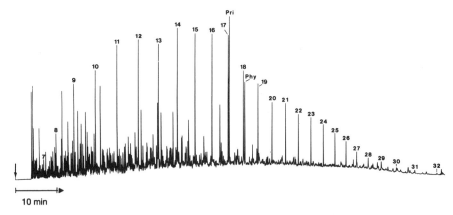

Figure 7.8 Use of thermal desorption for the analysis of trace volatiles in a rock sample. Column: 50 m × 0.32 mm FSOT coated with SE54 0.2 μm); cold trap: 0.5 mm diameter fused silica coated with 5 μm layer of polymethylsiloxane; desorption conditions: flow rate 5 ml min^{-1}; desorption temperature: 300 °C; cold trap: −120 °C; injection: 200 °C; column: 30 to 280 °C at 3 °C min^{-1}; detector: FID; carrier gas: helium. [Reproduced by permission of Huethig, from R. G. Schaefer, *HRC&CC*, **8**, 267 (1985)]

In the thermal desorption mode, a tube containing material from which compounds are to be desorbed is placed in the carrier gas flow line. This material may be an adsorbent such as Tenax TA on which volatiles have been collected previously off-line, or any sample in which the volatiles are required. Thus, powdered solid samples such as rock, polymers, paper, etc. have been examined by this technique. The tube is electrically heated to a suitable temperature to volatilize the volatiles and the gas flow carries them to a cold trap similar to that previously described in Figure 7.1 for SFE/GC. Other commercial systems use variations of the method for collecting the desorbed volatiles, including re-adsorption in a small trap near the start of the column. Thermal desorption techniques are used widely for the analysis of trapped atmospheric volatiles and headspace samples, while purge and trap techniques are favoured for aqueous samples. An example of the use of thermal desorption is given in Figure 7.8 and a purge and trap application is shown in Figure 7.9.

7.8 SAMPLE DERIVATIZATION

Derivatization is used in gas chromatography for the following reasons.

(a) To convert labile compounds to more thermally stable derivatives.
(b) To eliminate peak tailing with polar compounds

SAMPLE PREPARATION

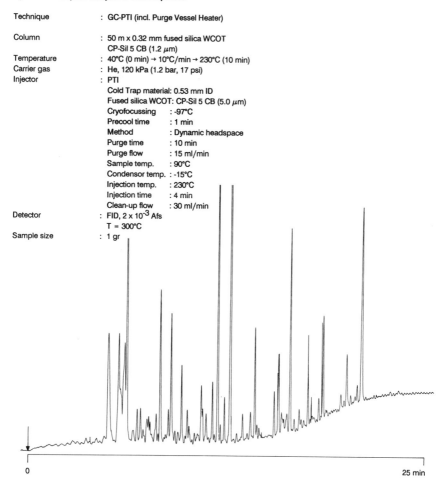

Figure 7.9 Use of purge and trap for fingerprinting of volatiles present in pillow down. (Reproduced by permission Chrompack International BV)

(c) To improve detectability, particularly of trace components in a complex matrix.

One of the limitations of gas chromatography is that the sample components must be thermally stable at the temperatures used by the instrument. Vaporization injection, in particular, imposes a more severe restriction as the injection temperature is nearly always higher than that of the column. Labile compounds, however, can sometimes be converted to more stable derivatives and so (a) is, in fact, the most common reason for choosing the option of derivatization.

The problem of peak tailing (b) usually arises as a result of the interaction of polar compounds with the column wall. This problem can often be solved by choosing an appropriate column for the analysis. This is usually a better option than derivatization, particularly where the removal of polar groups reduces the possibility of selective interactions with the stationary phase which can be helpful in separating difficult compounds. There are many commercial columns for the separation without peak tailing of strongly polar compounds such as amines, carboxylic acids, glycols and amides.

Compounds that often require derivatization include carbohydrates, polyols, thiols, polycarboxylic acids, high molecular weight alcohols and acids which may be labile at the high temperatures necessary for their separation, and polyfunctional compounds such as amino acids.

If it is decided to derivatize the sample then it is important for the chosen method to be rapid, quantitative and 'clean', i.e. the final sample should not produce interfering peaks as a result of the reaction, or from the reagent itself. For situation (b) where derivatization is applied as a way of enhancing the detection of trace components of complex mixtures, the object is to implant one or more halogen atoms in the molecule for electron capture detection. Sometimes the objective is to introduce a nitrogen-containing group for selective detection by NPD. Whatever the objective, commercial reagents are available to ensure that the above requirements are generally satisfied.

The majority of derivatization techniques for GC are based on silylation, alkylation, and acylation of active hydrogen atoms in hydroxylic compounds, carboxylic acids, amines and amides. The efficiency and ease of the derivatization procedure depends on the nature and configuration of the functional groups and on the reagent chosen. Silylation is usually the easiest technique to perform and, in most cases, silyl ethers or esters are formed very rapidly, quantitatively and usually at room temperature. Thus, the reaction can normally be carried out very simply in a small reaction vial simply by the addition of the reagent to the prepared sample. The conversion of hydroxylic compounds such as alcohols, carboxylic acids, phenols, sterols and carbohydrates to their trimethylsilyl ethers (or esters in the case of a carboxylic acid) can be described as follows.

$$ROH + (CH_3)_3SiR' = ROSi(CH_3)_3 + R'H$$

Similar reactions can be applied to amines, namely,

$$RNH_2 + (CH_3)_3Si R' = RNHSi(CH_3)_3 + R'H$$

or

$$RNH_2 + 2(CH_3)_3SiR' = RN(Si(CH_3)_3)_2 + 2R'H$$

7.8.1 Silylation Reagents

Trimethylchlorosilane $(CH_3)_3SiCl$

Trimethylchlorosilane (TMCS) is the simplest form of silylation reagent but it is rarely used today by itself, partly because of its noxious nature, but mainly because it is a poor silyl donor. It is sometimes added to accelerate silylation with other reagents. When it is chosen, a base solvent such as pyridine or diethylamine is essential for the reaction to proceed.

Hexamethyldisilazane $(CH_3)_3SiNHSi(CH_3)_3$

Hexamethyldisilazane (HMDS) was an early replacement for TMCS and is effective for the trimethylsilylation of phenols [24] and many simple hydroxylic compounds [25]. Nevertheless, a range of more effective silyl donors are now available and so HMDS is rarely used.

Bis(N-O-trimethylsilyl)acetamide $CH_3C[OSi(CH_3)_3] = N[Si(CH_3)_3]$

Bis(N-O-trimethylsilyl)acetamide (BSA) reagent is used extensively for derivatizing alcohols, amines, carboxylic acids, phenols, steroids, biogenic amines and alkaloids, but it is not recommended for carbohydrates. Although the reaction is nearly always fast, it is used usually in conjunction with pyridine or DMF, particularly for phenols. Both N- and O- derivatives are formed with amino acids, and N-TMS derivatives are formed from aromatic amides [26].

*N,O-bis(trimethylsilyl)trifluoroacetamide
$CF_3C[OSi(CH_3)_3] = N[Si(CH_3)_3]$*

Replacement of the terminal methyl hydrogens by fluorine gives N,O-bis(trimethylsilyl)trifluoroacetamide (BSTFA) which again is a powerful silyl donor. Its main advantage is the greater volatility of the reaction byproducts which ensures that they elute before volatile derivatives, particularly of compounds such as lower amino acids.

*N-Methyl-n-trimethylsilyltrifluoroacetamide
$CF_3(CO)N(CH_3)[Si(CH_3)_3]$*

N-Methyl-n-trimethylsilyltrifluoroacetamide (MSTFA) is more volatile than either BSTFA or BSA, giving even more volatile byproducts. The reagent is particularly useful for steroids and amine hydrochlorides.

N-Trimethylsilyldiethylamine $Si(CH_3)_3N(C_2H_5)_2$

N-Trimethylsilyldiethylamine (TMSDEA), a strongly basic silyl donor, is particularly useful for derivatizing low molecular weight acids, and the byproduct, diethylamine can be easily removed by gassing from the reaction mixture. TMSDEA also reacts with amino acids, amines and, of course, hydroxylic compounds.

N-Trimethylsilylimidazole $Si(CH_3)_3(C_3N_2H_3)$

N-Trimethylsilylimidazole (TMSI) is generally regarded as the strongest silyl donor available, and is the reagent of choice for carbohydrates [27] and steroids, including those which are highly hindered. It is also used for the derivatization of alcohols, phenols, organic acids, hormones. glycols, nucleotides and narcotics. A unique feature of this reagent is that it does not react with amino groups.

N-Methyl-N-(tert-butyldimethylsilyl)-trifluoroacetamide $CF_3(CO)N(CH_3)Si(C_4H_9)(CH_3)_2$

N-Methyl-*N*-(*tert*-butyldimethylsilyl)-trifluoroacetamide (MTBSTFA) is the tertiary butyl homologue of MSTFA and gives more stable dimethyl-*tert*-butyl silyl ethers of hydroxyl, carboxyl and thiol compounds, as well as primary and secondary amines.

The introduction of halogen atoms into the silylated derivative will enhance its electron capture detectability and this is a very useful way of determining trace amounts of certain types of compound in complex matrices. One type of derivative for this purpose replaces one of the TMS methyl groups with a pentaflurophenyl group. These pentafluorophenyldimethylsilylethers (flophemesyl derivatives) can be prepared from flophemesyl chloride or bromide. This technique has not found widespread popularity as a means of trace analysis, largely because of the excessive amounts of highly halogenated byproducts which can interfere easily with the final separation, and the relatively unstable nature of the halogenated derivatives.

7.8.2 Alkylation Methods

Alkylation is used extensively for derivatizing higher fatty acids before their separation, although other types of compound may be treated similarly in appropriate cases. Alkylation methods are based on the following.

1. Diazomethane generated by the reaction of sodium hydroxide on *N*-methyl-*N'*-nitro-*N*-nitrosoguanidine ether in a special microgenerator. This technique is used mainly for higher fatty acids but care should be taken due to the toxic and potentially explosive nature of diazomethane.

SAMPLE PREPARATION

2. Boron trifluoride/methanol, also for fatty acids.
3. Dimethylacetal with dimethylformamide as solvent methylates higher carboxylic acids and phenols.
4. Pentafluorobenzylbromide (PFB Br) has been recommended by Kawahara for the trace analysis of mercaptans, phenols and organic acids in surface water by electron capture detection [28,29]. It can also be used for carboxylic acids, phenols and sulphonamides. The method relies on the use of a tetrabutylammonium counter ion and methylene chloride as solvent.

Alkylation usually proceeds quantitatively and quickly in most cases, although perhaps not as easily as silylation. Usually some heating of the reaction mixture is required for the technique to be completely effective.

7.8.3 Acylation Techniques

Acylation is often a preferred method for the derivatization of amines, alcohols and phenols, when selective detection is required by ECD. Commercial reagents for this purpose include the following.

1. Fluorinated acid anhydrides. The anhydrides of trifluoracetic acid (TFAA) pentafluoropropionic acid (PFPA) and heptafluorobutyric acid (HFBA) react readily with alcohols, phenols and amines to form stable fluoroacyl

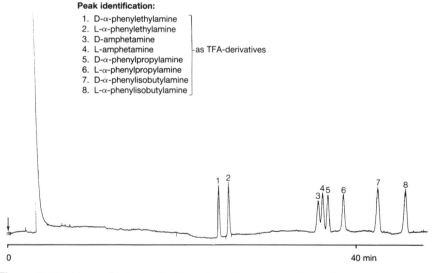

Figure 7.10 Use of derivatization for the separation of enantiomers of amphetamine and other amines on a chiral stationary phase. Column: 25 m × 0.22 mm FSOT column coated with Chirasil-L-Val (0.12 μm); temperature: 100 °C; nitrogen carrier gas; split injection. (Reproduced by permission of Chrompack International BV)

derivatives for both FID and ECD. Direct acylation is possible but a base solvent such as triethylamine is recommended as an acid acceptor to enhance reactivity. Figure 7.10 shows the use of acylation for the separation of amphetamine enantiomers and other optically active primary amines using a chiral stationary phase.

2. Fluoroacylimidazoles. These reagents offer some advantages over the use of acid anhydrides as no acids are released into the reaction mixture, and the acylated products are quite volatile. The trifluoromethyl (TFAI), pentafluoropropyl (PFPI) and heptafluorobutyl (HFBI) imidazoles react readily with hydroxyl groups and primary and secondary amines.

Table 7.4 Derivatization reagents and their application

Functional group	General detection (FID) Reaction type	Reagent	Selective detection (ECD) Reaction type	Reagent
Carboxylic acids	Silylation	BSTFA MTBSTFA	Silylation	Flophemesyl chloride
	Alkylation	Diazomethane BF$_3$methanol dimethylacetal	Alkylation	PFB Br
Hydroxy groups:				
Primary/secondary Alcohols	Silylation	TMSI (best reagent if water present)	Silylation	Flophemesyl chloride
Polyols	MSTFA			
Carbohydrates	MTBSTFA			
Thiols				
	Alkylation	BF$_3$/methanol	Alkylation	PFB Br
Phenols	Silylation	BSTFA TMSI	Silylation	Flophemesyl chloride
	Alkylation	Dimethylacetal	Acylation Alkyloation	HFBA PFB Cl
Carboxyl groups	Oxime	Methoxy-amine	Oxime Hydrazone	PFBHA 2,4-Dinitrophenylhydrazine
Amines (general)	Silylation	MSTFA for GCMS of basic drugs		
Primary	Acylation	TFAA PFPA	Alkylation Acylation	PFB Cl HFBA
	Silylation	MBTSTFA		HFBI MBTFA
Secondary	Acylation	TFAA PFPA	Alkylation Acylation Carbamate	PFB Cl HFBI Trichloroethyl Chloroformate
(Barbiturates)	Alkylation	TMAH		
Tertiary			Carbamate	Trichloroethyl Chloroformate

Reproduced by permission of Chrompack International BV.

SAMPLE PREPARATION

3. *N*-Methyl-bis(trifluoroactamide) (MBTFA). MBTFA is a trifluoracetylating reagent for primary and secondary amines, hydroxyl and thiol groups under mild and non-acidic conditions. It also acylates amines selectively in the presence of hydroxyl and carboxyl groups. The conversion of sugars to their trifluoroacetyl derivatives with this reagent gives volatile products and quantitative results [30].

Numerous procedures have been developed for specific types of compound, and the books by Pierce [31] and Knapp [32] present the most comprehensive coverage available to date on this subject.

A summary of derivatization recommendations is given in Table 7.4.

REFERENCES

1. Grob, R. L. and Kaiser, M. A. (1982) *Environmental Problem Solving using Gas and Liquid Chromatography*, Journal of Chromatography Library Vol. 21, Elsevier, Amsterdam.
2. Dix, K. D. and Fritz, J. S. (1987) *J. Chromatogr.*, **408**, 201.
3. Dix, K. D. and Fritz, J. S. (1990) *Anal.Chim.Acta*, **236**, 43.
4. Godefroot, M., Sandra, P. and Verzele, M. (1981) *J. Chromatogr.*, **203**, 325.
5. Poole, S. K., Dean, T. A. and Poole, C. F. (1987) *J. Chromatogr.*, **400**, 323.
6. Kepner, R. E., van Straten, S. and Weurman, C. (1969) *J. Agric. Food Chem.*, **17**, 1123.
7. Robins, W. R. (1979) *Anal.Chem.*, **51**, 1860.
8. Schill, G., Ehrsson, H., Vessman, J. and Westerlund, D. (1984) *Separation Methods for Drugs and Related Organic Compounds*, 2nd edn, Swedish Pharmaceutical Press, Stockholm, Sweden.
9. Davies, I. L., Rayner, M. W., Kithinji, J. P., and Bartle, K. D. (1988) *Anal. Chem.*, **60**, 683A.
10. Lee, M. L. and Markides, K. E. (eds) (1990) *Analytical Supercritical Fluid Chromatography and Extraction*, Chromatography Conferences Inc., Provo, UT.
11. McDowall, R. D., Pearce, J. C. and Murkitt, G. S. (1989) *Trends. Anal. Chem.*, **8**, 134.
12. Liska, I., Krupcik, J. and Leclerq, P. A. (1989) *J. High Resn Chromatogr.*, **12**, 577.
13. Dressler, M. (1979) *J. Chromatogr.*, **165**, 167.
14. Cortes, H.J. (ed.) (1990) *Multidimensional Chromatography: Techniques and Applications*, Dekker, New York.
15. Grob, K. and Schilling, B. (1985) *J. High Resoln Chromatogr.*, **8**, 726.
16. Munari, F. and Grob, K. (1990) *J. Chromatogr. Sci.*, **28**, 61.
17. Hachenberg, H. and Schmidt, A. P. (1977) *Gas Chromatographic Headspace Analysis*, Heyden, London.
18. Kolb, B. (1980) *Applied Headspace Gas Chromatography*, Heyden, London.
19. Schreier, P. (1984) *Analysis of Volatiles: Methods and Applications*, Walter de Gruyter, Berlin.
20. Later, D. W., Wilson, B. W. and Lee, M. L. (1985) *Anal.Chem.*, **57**, 2979.
21. Schuetzle, D., Rilet, T. L., Prater, T. J., Harver, T. M. and Hunt, D. F. (1982) *Anal. Chem.*, **54**, 265.

22. Hagen, D. F., Markell, C. G., Schmitt, G. A. and Blevins, D. B. (1990) *Anal. Chim. Acta.*, **236**, 157.
23. Brouwer, E. R., Lingeman, H. and Th. Brinkman, U. A. (1990) *Chromatographia*, **29**, 415.
24. Grant, D. W. and Vaughan, G. A. (1962) in *Gas Chromatography 1962*, (ed. M. van Swaay), Butterworths, London, pp. 304–314.
25. Sweeley, C. C., Bentley, A., Makita, M. and Wells, W.W. (1963) *J. Amer. Chem. Soc.*, **85**, 2497.
26. Klebe, J. F., Finkbeiner, H. and White, D. M. (1966) *J. Amer. Chem. Soc.*, **88**, 3390.
27. Brittain, G.D. and Schewe, L. (1971) *Recent Advances in Gas Chromatography*, (eds I. I. Domsky and J. A. Perry), Marcel Dekker, New York.
28. Kawahara, F. K. (1968) *Anal. Chem.*, **40**, 1009.
29. Kawahara, F. K. (1968) *Anal. Chem.*, **40**, 2073.
30. Sullivan, J. and Schewe, L. (1977) *J. Chromatogr. Sci.*, **15**, 196.
31. Pierce, A. E. (1979) *The Silylation of Organic Compounds*, Pierce Publisher.
32. Knapp, D. (ed.) (1979), *Handbook of Analytical Derivatization Reactions*, Wiley, New York.

8
Analysis and Optimization

8.1 INTRODUCTION

In previous chapters of this book a considerable emphasis has been placed on the simplicity of the open tubular column geometry, and the practical usefulness of the Golay theory. It is often argued by users of gas chromatography that theory is 'irrelevant to their needs', particularly when they are not involved in method development. Many are only interested in the successful application of specific test methods or perhaps in producing an effective separation of mixtures obtained in the course of research studies. Such needs without the necessity of an in-depth knowledge of the subject must, of course, be regarded sympathetically. Capillary GC is an analytical tool applied and supported by sophisticated equipment and data systems, and it must be capable of being used for everyday purposes without the need for extensive expertise by the user. This chapter is written primarily from this point of view. It includes practical guidance on methods of quantitation, adjustment of isothermal and programmed conditions and trace analysis. **The overall objective is to attain the required degree of resolution between critical pairs inside a maximum time limit and to produce a sufficient response from the detector to give the required degree of accuracy and precision.**

8.2 QUANTITATIVE ANALYSIS IN CAPILLARY GC

The accuracy and precision of quantitative analysis in GC have been subjects of considerable investigation, and are still highly controversial topics. Perhaps it is not surprising that estimates of the accuracy of the technique vary widely in view of the number and variety of the variables. Thus errors can arise from numerous sources, particularly in sample preparation and handling, the injection procedure, the amount of resolution, the detector and finally the handling and processing of the data.

These features have all been discussed in previous chapters but the operator should always be aware of the very small scale of the method and should take particular care when handling and injecting the sample. In fact, sample introduction almost certainly accounts for the majority of the analytical error.

Sample preparation errors cannot, of course, be considered as part of the intrinsic error of capillary GC, although any estimation of the total analytical error should include this aspect. The degree of error will be very specific to the application. In considering the errors involved in the chromatographic part of the overall procedure it is assumed that the off-line techniques are carried out effectively and that errors in any standardized procedure have been fully evaluated during the method development stage.

8.2.1 Gas Analysis

Gas analysis is certainly the most accurate application in gas chromatography because gas volumes can be sampled accurately using calibrated gas sample valves, and there is no possibility of discrimination errors because of the absence of a vaporization stage. Although the chromatographic literature is now very extensive, there is comparatively little information on general methods of gas analysis by capillary GC. The books by Jeffery and Kipping and, more recently, Cowper and DeRose still represent the state-of-the-art in this area [1, 2].

The **external standard** method is normally employed for gas analysis. This involves calibrating the GC system with standard mixtures of the relevant components to determine response factors (see Section 3.8.4). The same volume of the unknown sample is then analysed by chromatography using identical conditions, and the concentrations of the components calculated from the peak areas and the response factors. Figure 8.1a illustrates the principle of the external standard method.

A minimum of three calibration gas mixtures containing the components of interest should be analysed initially to establish the linearity of the system, and these should be admitted to the column by means of a gas sample valve and calibrated sample loop. For normal bore capillary columns the sample volume is usually 1–2 ml using split injection with a split ratio of about 100:1. For wide bore columns using direct injection, the sample volume should be no higher than 100–200 μl.

The peak areas are measured and response factors for each component calculated as follows:

$$RF_i = \frac{A_i}{c_i} \tag{8.1}$$

where RF_i = response factor for component i; A_i = peak area (arbitrary units) given by component i; and c_i = concentration of component i in the original standard mixture.

Provided that the same sample loop is used for the calibration and analytical runs then its actual volume is unimportant and so does not need to be known accurately. An alternative method would be to use the same calibration sample but to analyse various volumes using several sample loops, and then to calculate response factors with respect to the actual mass of each component analysed. This method is not recommended, however, because the calibrated volume of every loop would have to be known accurately, and it would involve the inconvenience of having to change to sample loop for every run.

For constant volume sampling, the response factors should all agree within the required limits of accuracy. Any disagreement implies that the detector is non-linear or that the integrator setting is producing bias errors. The micro-TCD is normally used for permanent gas analysis with capillary columns, and this detector should give a linear response provided it is operated according to the recommendations given in Chapter 3, i.e. with hydrogen or helium carrier gas. The FID would preferably be used for gaseous hydrocarbons and gives a linear response towards most compounds.

Once linearity has been established it should then be sufficient to carry out subsequent calibrations with a single reference mixture containing concentrations of components at about the same level as those in the samples.

After calibration the analysis is performed simply by repeating the same injection procedure with the unknown sample. Concentrations are then calculated from

$$c'_i = \frac{A'_i}{RF_i} \qquad (8.2)$$

$$= \frac{A'_i}{A_i} \times c_i \qquad (8.3)$$

where c'_i and A'_i now refer to the unknown sample. For maximum accuracy, each analysis and calibration run should be carried out alternately.

Barratt has reviewed methods for the preparation of standard gas mixtures including sections on fundamental principles, and static and dynamic mixing methods [3]. Calibration standards, including mixtures of permanent and hydrocarbon gases are available from most commercial suppliers. The accuracy of gas analysis by capillary GC should lie in the range 0.1–0.5% coefficient of variation.

8.2.2 Analysis of Liquid Samples

In capillary GC most quantitative analysis is performed on liquid samples for the following reasons.

1. To convert a solid sample to the appropriate state for introduction to the chromatograph by syringe techniques.
2. To dilute any sample to an acceptable level so as not to exceed the column capacity.
3. To ensure that no involatile solids are present
4. The sample preparation technique terminates in a liquid sample.
5. This is the natural state of the sample, as for instance in the case of many aqueous samples from environmental sources.

Considerable care should be exercised in the selection of solvents: initial tests should be carried out to ensure that there are no interfering impurities, and there must be no overlap between the solvent peak and any of the required components. Volatile solvents are normally selected which elute close to the holdup time of the column and so it is nearly always the most volatile component. The solvent should not be **excessively** volatile, however, as this can give rise to peak tailing as a result of back-diffusion of vapour in the injector, or premature vaporization before the sample has completely penetrated the septum. Hexane, heptane or, for more polar samples, C_2–C_3 alcohols are preferred over very volatile solvents such as acetone, tetrahydrofuran (THF), pentane, methylene chloride or methanol.

The basic external standard procedure, described in Section 8.2, for the analysis of gases is generally unsuitable for liquids because of their much smaller injection volumes. As pointed out in Chapter 6, typical sample volumes for capillary columns range from about 0.1 μl for split injection up to several microlitres for on-column injection. It was seen that a variety of factors can affect the accuracy of these volumes and even with the most meticulous care the coefficient of variation for sample volume is considerably greater than 1% for manual injections. Automatic injection with autosamplers will be somewhat better than this due to the exact duplication of the procedure, but even here the use of the external standard method is still not sufficiently accurate for most purposes. There may be an exception in some cases of trace analysis where relative errors at the percentage level become unimportant. For instance, a variation of ±10% at 1 ppm gives a range of 0.9–1.1 ppm.

To overcome the problem of sample volume variation, the **internal standard** method as illustrated in Figure 8.1b is the most accurate of the various methods used for quantitative analysis in capillary GC. This method involves adding a known concentration of a suitable reference compound to the sample before injection, and the **ratios** of the component peak areas to that of the reference compound are calculated. These ratios are independent of the amount of sample injected because all peak areas will vary in the same proportion with changes in the sample size provided that the detector response is linear. The added reference compound is called the internal standard and is chosen wherever possible to be of the same chemical type as the sample

ANALYSIS AND OPTIMIZATION

components, although this is not always feasible. A more important property is that the standard is pure and gives a peak in a vacant region of the chromatogram. Also, it is advisable to avoid excessive differences in retention time between the internal standard peak and the sample components so as to minimize analytical error. In some cases, particularly in the programmed temperature analysis of very complex samples, it may even be advantageous to use more than one standard.

To apply the internal standard method, **relative response factors**, RRF are first determined by preparing several calibration mixtures containing known concentrations of each of the pure components and the internal standard. The calibration mixtures are injected, peak areas are measured and relative response factors calculated from:

$$RRF_i = \frac{A_i \times c_s}{A_s \times c_i} \tag{8.4}$$

where c_s and A_s are the concentration and peak area of the standard compound respectively. A_i and c_i are the same properties for the sample components as in the external standard method, (8.3).

Analysis is performed by adding a known concentration, c'_s, of the same standard to the unknown sample and then analysing it under identical conditions as used for calibration.

The concentration of each component c'_i is then determined from,

$$c'_i = \frac{A'_i \times c'_s}{A'_s \times RRF_i} \tag{8.5}$$

where the apostrophized values relate to the unknown sample being analysed.

Other ways of minimizing errors in the internal standard procedure are listed as follows.

- Always use the same concentration of standard by adding it to the sample volumetrically.
- Keep the sample volume nominally constant for all runs.
- Recalibrate frequently and certainly after any change of column or conditions.

It can often be impossible to find vacant areas in a complex chromatogram in which to situate an internal standard. In such cases it is possible to choose an occupied area and to make the appropriate corrections. It is not necessary for the component peak which occurs at the chosen position to be the same as the internal standard but it is very important that the added standard increases the peak area by a significant amount as shown in Figure 8.1c. As the **corrected internal standard** method also gives the area of the peak occurring under the standard peak it can be used to determine this component as well. In fact, this is frequently the main objective of the analysis in which case the added

(a) External standard

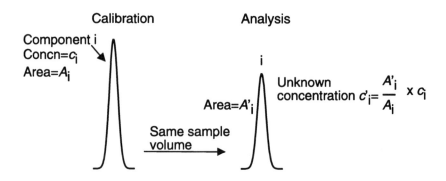

(b) Internal standard (I.S)

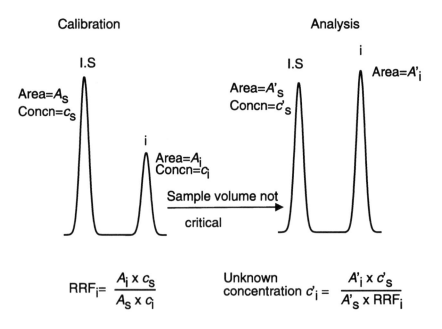

Figure 8.1 Quantitative analytical techniques in capillary GC

ANALYSIS AND OPTIMIZATION

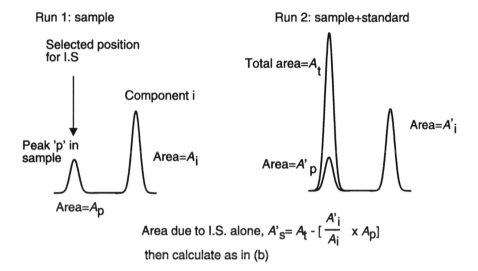

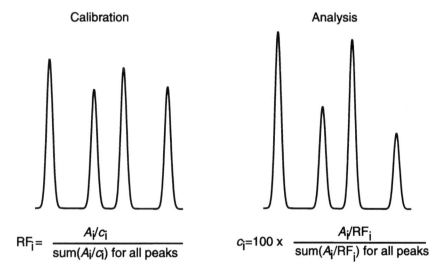

Figure 8.1 (*continued*)

standard **will** have the same identity and the method is then referred to as **spiking**.

The **internal normalization** technique is sometimes employed and this is illustrated by Figure 8.1d. This method is only valid when all components of the sample appear on the chromatogram and all of their areas are measured. The areas are corrected for the response factors measured from calibration runs, determined as follows:

$$RF_i = \frac{A_i}{c_i \sum \left(\frac{A_i}{c_i}\right)} \qquad (8.6)$$

where $\sum_1^n A_i$ is the sum of all component areas on the chromatogram (n peaks). Subsequent analysis is performed by running the unknown sample and calculating each component from,

$$c_i' = \frac{A_i'}{RF_i \sum \left(\frac{A_i'}{RF_i}\right)} \qquad (8.7)$$

The internal normalization technique can only be applied reasonably to very simple samples such as fractions from distillation. Remember that water does not give a response in the FID and so considerable caution must be exercised before deciding to use this method.

Although areas are usually used for the calculations of component concentrations, peak heights are also sometimes used, particularly when there is some interference between the peaks resulting in troughs rather than a complete return to the baseline. Thus, peak heights are affected far less by a limited amount of peak overlap than integrated peak areas. **This option should not, however, be used as a way of accepting poor peak resolution as there is no substitute for good chromatographic practice when the goal is to achieve the best quantitative results.**

8.3 PRINCIPLES AND PRACTICE OF TRACE ANALYSIS

Trace analysis is an important application field for capillary GC and covers a wide variety of disciplines although environmental monitoring of atmospheres, effluents, soils, etc., is of particular importance nowadays. Many of these problems are solved using various forms of headspace, thermal desorption, and purge and trap methods which were discussed in Chapter 7. In fact, in any situation where the concentrations are known to be at the ppb level or less then special sample preparation techniques are essential to increase the concentrations to measurable levels.

There is no need to define exactly what is meant by trace analysis but the

ANALYSIS AND OPTIMIZATION

user will be aware of many situations in which the components are not easily seen when they are applied to the column in the usual way. A loose definition is the quantitative measurement of components at sub-percentage levels. Unfortunately, common reactions to the inability to see component peaks are to inject increasing volumes of the sample; to decrease the detector attenuation to the point at which a very noisy baseline is obtained; or to carry out quite irrelevant adjustment to the equipment.

Trace analysis should be approached in a systematic fashion. Firstly, the column should be running under conditions which produce peaks at known retention times as determined by preliminary runs with standard mixtures. Minimum detectable limits should then be determined (see Section 3.8.3) for these compounds to assess the feasibility of performing the analysis without any further sample treatment. Minimum detectabilities are critical with respect to trace analysis and are affected by the choice of column and its operating conditions. Understanding these variables, therefore, can help considerably in extending the detectability of the equipment.

8.3.1 Problems of Trace Analysis

Two major types of sample are involved in trace analysis after they have been prepared for sample injection.

(a) The major component is solvent, usually water. The components to be measured are all in trace quantities and so sufficient sample volume must be introduced to the column to enable the detector to give measurable responses. If an organic solvent is the major component, the sample can usually be concentrated by one of the conventional sample preparation techniques. Figure 8.2a shows an example of this type of analysis, but an organic solvent has been used. The sample consists of a range of derivatized steroids at concentration levels of $20 \, \text{ng} \, \mu l^{-1}$ and was introduced to the column by splitless injection.

(b) There are major components distributed among the trace components which cause interference on the chromatogram, particularly when they overload the column to give fronting or tailing peaks. This will usually occur if the sample size is large enough for the minor components to be seen. An example of a type (b) sample is shown in Figure 8.2b. Note that the methane concentration level of 5 ppm is sufficient to enable it to be detected and measured using normal split injection from a gas sample of $150 \, \mu l$.

Type (a) samples are the easiest to cope with as they simply require the use of either splitless or on-column injection to enable a larger sample volume to be applied to the column. The amounts of the trace components must, however, be above the minimum detectability of the system. These injection methods would normally also be used for type (b) samples but although the

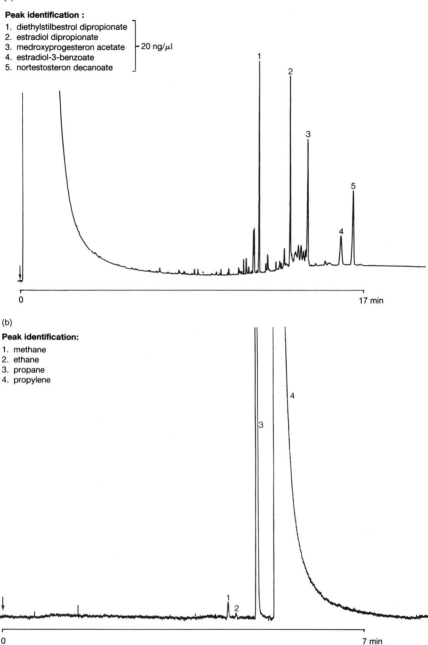

Figure 8.2 Types of trace analysis. (a) Determination of trace steroids. Column: 3 m × 0.1 mm (0.2 μm) coated with CPsi19CB; temperature 70 to 300 °C at 20 °C min^{-1}; splitless injection of 1.2 μl solution containing 20 ng μl^{-1} of each component; detector: FID; hydrogen carrier gas. (b) Determination of impurities in propylene. Column: 50 m × 0.32 mm FSOT PLOT alumina/KCl; temperature: 130 °C; nitrogen carrier gas; split injection of 150 μl gas; methane concentration = 5 ppm. (Reproduced by permission of Chrompack International BV)

peak areas for the minor components may be sufficiently large in standard mixtures, a satisfactory resolution may not be obtained with real samples due to the interference problem from major components. For rapidly eluting components such as the gases shown in Figure 8.2b, normal split injection may produce satisfactory results. Of course, effective method development should have ensured that all components are resolved sufficiently at normal levels of concentration, and so if interference is experienced then it would be due to overloading. If the latter is the case then the only solution is to increase the minimum detectability of the system so that smaller sample sizes can be applied.

8.3.2 Factors Affecting Instrument Sensitivity

The sensitivity of the instrument depends critically on the type of detector, the column and its operating conditions. As we have seen in Chapter 3, there are two types of detector for GC: mass flow sensitive, typified by the FID, and concentration sensitive detectors, typified by the TCD or ECD. Table 8.1 gives an indication of the minimum detectabilities for each type of detector calculated from commercial data and assuming the use of a 25 m × 0.25 mm i.d. column and a retention time of 20 minutes. These figures are given for rough guidance only as there may well be considerable divergences in practice. Minimum detectabilites depend on the stability of the chromatograph under the conditions used and will vary from one system to another. They also depend critically on column bleed levels. Also, data given by many commercial manufacturers give 'minimum detectabilities' without any operating data. Such data may refer, in fact, to the 'detectivity', which is the minimum detectable amount without a column connected, perhaps on a specialized test rig and so does not include the effect of column dilution. The data in Table 8.1 has been calculated for a constant sample addition size of 1 μl but this can be increased, particularly in on-column injection methods, thereby decreasing the minimum detectability.

Equations (3.12) and (3.14) describe the peak maxima in terms of concentration and mass flow per second respectively. These relationships are based on isothermal operation, however, and a more general relationship is required to express the detector response in a way that is independent of the column temperature operating conditions.

The Golay theory shows that the plate number depends on the choice of compound and its retention factor, as well as a number of other variables which depend on column temperature. Nevertheless, columns are usually characterized by a generalized plate number which, for efficient columns, has a value close to the ratio of the column length to its internal diameter. The plate number does not, in fact, vary dramatically when it is calculated on different peaks in the chromatogram and so we shall assume that the plate number is constant and independent of temperature for the following

Table 8.1 Minimum detectabilities. (Calculated for a 26 m × 25 mm column. Assumed retention time of 20 minutes and helium carrier gas at 25 cm^{-2}

Detector	Injection method Assuming 1 μl sample	Minimum detectability
FID	Split (100:1)	1–10 ppm
	Splitless	10–100 ppb
	On-column	10–100 ppb
NPD	Split(100:1)	N: 2–20 ppm
		P: 1–5 ppm
	Splitless	N: 20–100 ppb
		P: 20–100 ppb
	On-column	N: 20–100 ppb
		P: 20–100 ppb
FPD	Split(100:1)	S: 100–1000 ppm
		P: 1–10 ppm
	Splitless	S: 1–10 ppm
		P: 10–100 ppb
	On-column	S: 1–10 ppm
		P: 10–100 ppb
ECD (lindane)	Split(100:1)	10 ppb–100 ppm
	Splitless	100 ppt–1 ppb
	On-column	100 ppt–1 ppb
TCD	Split(100:1) (splitless and on-column not normally used with TCD)	100–1000 ppm

treatment. This means that the length of the column occupied by all the chromatographic zones immediately prior to their elution is constant and so the heights and widths of peaks will depend solely on the respective rates at which the zones elute from the column.

The plate theory of chromatography shows that the mass of solute in the final plate of the column when the zone maximum passes through it is given by equation (2.35), namely,

$$Q_r(\max) = \frac{w_i}{\sqrt{2\pi N}} \tag{8.8}$$

The length of the last plate is L/N (i.e., the plate height) and on exit, each zone is travelling at a velocity of $u/(1 + k_T)$ cm s^{-1}, where k_T is the retention factor of the solute at the column temperature at the elution time. Under programming conditions, this temperature is known as the **retention temperature** and the time taken for this zone of maximum concentration to vacate the

ANALYSIS AND OPTIMIZATION

column is $L(1 + k_T)/Nu$. If the volumetric carrier gas flow rate is F_V then the volume of carrier gas required to elute the final plate contents is $F_V L(1 + k)/Nu$.

Thus,

maximum concentration of zone in carrier gas = $\dfrac{\text{mass in last plate at maximum}}{\text{volume of carrier gas required to elute it}}$

i.e.,

$$C_{max} = \frac{Q_r(max)Nu}{F_V L(1 + k_T)} \quad (8.9)$$

$$= \frac{W_i}{v_0(1 + k_T)} \sqrt{\frac{N}{2\pi}} \quad (8.10)$$

where v_0 is the geometric volume of the column. This equation is of course similar to (3.11) except that it specifically refers to the column exit conditions, particularly the retention factor k_T at the retention temperature.

The detector response, R_D, for concentration dependent detectors is proportional to C_{max} where the proportionality constant is the detector sensitivity, S, i.e.,

$$R_D(\text{conc}) = \frac{w_i S}{v_0(1 + k_T)} \sqrt{\frac{N}{2\pi}} \quad (8.11)$$

but

$$v_0 = \frac{\pi d_i^2 L}{4} \quad (8.12)$$

Therefore

$$R_D(\text{conc}) = \frac{4 W_i S}{\pi d_i^2 L(1 + k_T)} \sqrt{\frac{N}{2\pi}} \quad (8.13)$$

and for mass flow sensitive (mfs) detectors

$$R_D(\text{mfs}) = S \times F_v \times C_{max} \quad (8.14)$$

$$= \frac{S F_v W_i}{V_0(1 + k_T)} \sqrt{\frac{N}{2\pi}} \quad (8.15)$$

$$= \frac{S u W_i}{L(1 + k_T)} \sqrt{\frac{N}{2\pi}} \quad (8.16)$$

If w_i represents a trace component, then it will be limited by the maximum volume of liquid sample that can be analysed. This maximum volume, $I_s(\text{max})$, depends on the method of sample injection, the presence or otherwise of interfering major components [type (b) samples] and on the volume capacity (Chapter 4). Volume capacity, however, is only important under isothermal column conditions because programmed temperature operation has the effect of focusing the sample during application hence minimizing the effect on the final peak width.

The detector response towards trace components can, therefore, be stated as

$$R_D(\text{conc}) = \frac{4Sc_i I_v(\text{max})}{\pi d_i^2 L(1 + k_T)} \sqrt{\frac{N}{2\pi}} \qquad (8.17)$$

$$= \frac{4Sc_i I_s(\text{max})}{\pi d_i^2 L\left(1 + \dfrac{4d_f K_T}{d_i}\right)} \sqrt{\frac{N}{2\pi}} \qquad (8.18)$$

and

$$R_D(\text{mfs}) = \frac{Suc_i I_s(\text{max})}{L(1 + k_T)} \sqrt{\frac{N}{2\pi}} \qquad (8.19)$$

$$= \frac{Suc_i I_s(\text{max})}{L\left(1 + \dfrac{4d_f K_T}{d_i}\right)} \sqrt{\frac{N}{2\pi}} \qquad (8.20)$$

Where K_T = the partition coefficient at retention temperature T and c_i is the concentration of any trace component i. For any trace analysis to be successful, it is important that the detector response is greater than twice the noise level. Thus the minimum detectability D is when $R_D = 2 \times$ noise level (see Chapter 3).

By examining these two equations we can see how to increase the response, and hence the minimum detectability for trace analysis.

8.3.3 Increasing Instrument Detectability

Maximum sample volume, $I_s(\text{max})$

The maximum sample volume that can be applied to the column without distortion of trace component peaks and interference by major components depends on the particular application and the nature of the sample components.

$I_s(\max)$ can be increased in proportion to the amount of stationary phase in the cross-sectional area and this is proportional to the product of the film thickness and column diameter.

Column diameter, d_i

For concentration detectors, (8.18) shows that an increase in d_i would increase the denominator and apparently **decrease** the detectability. However $I_s(\max)$ increases in proportion to d_i if the film thickness is constant and so the overall detectability remains almost constant. With mass flow detectors such as the FID, increasing the column diameter has a twofold advantage according to (8.20), because I_s is increased, thereby increasing the numerator, and the denominator is decreased. The overall effect is to increase the detectability by the **square** of the diameter approximately, assuming that the gas velocity is kept constant. In practice, the optimum velocity decreases with increasing diameter and so this would tend to have a moderating effect on detectability.

Stationary phase film thickness, d_f

It is often thought mistakenly that increasing the film thickness automatically increases the detectability of the chromatographic system in the same proportion. Certainly the maximum sample volume $I_s(\max)$ increases in direct proportion to d_f but whether or not this results in an enhancement of the peak heights depends on the column conditions.

If the retention factors are initially very low then there will be an immediate advantage in increasing the film thickness by virtue of the increased $k/(1+k)$ factor of the resolution equation. Choosing the appropriate film thickness should, however, be a normal part of the optimization routine during initial trials.

Equations (8.19) and (8.20) indicate that increasing the film thickness and increasing $I_s(\max)$ in proportion results in an almost constant detectability with any type of detector if K_T is constant, i.e. under isothermal conditions. Also, the analysis time would be extended due to the increased retention times of the components. In fact, the overall resolution could even be inferior to that on the previous column because of a lowered column plate number. **To gain the real advantage of using a thicker stationary phase film in trace analysis, it is essential to operate the column under programmed temperature conditions.**

The theory of programmed temperature operation (see Section 8.4) shows that retention temperature increases with film thickness if other conditions are constant. This means that the proportionate increase in the exit retention factor k_T for any specific compound is less than the increase in the film thickness. Hence, if the column is loaded to its maximum amount, i.e. $I_s(\max)$, then there will be an enhancement in the peak heights, or detectability. From (8.19) by taking the ratio of detectabilities for two columns coated with different film thickness,

$$\frac{\text{peak height on column 2}}{\text{peak height on column 1}} \cong \frac{I_s(\max)(2)}{I_s(\max)(1)} \times \frac{1+k_{T1}}{1+k_{T2}} \quad (8.21)$$

If we assume that the maximum sample loading is proportional to the film thickness, then

$$\frac{I_s(\max)(2)}{I_s(\max)(1)} = \frac{d_f(2)}{d_f(1)} \quad (8.22)$$

Hence

$$\frac{\text{peak height on column 2}}{\text{peak height on column 1}} = \frac{d_f(2)}{d_f(1)} \times \frac{1+k_{T1}}{1+k_{T2}} \quad (8.23)$$

Figure 8.3 shows plots of peak height versus film thickness calculated for a series of columns coated with film thicknesses of 0.1, 0.2, 0.5, 1, 2 and 5 μm. The calculations are based on the assumption that starting temperatures and

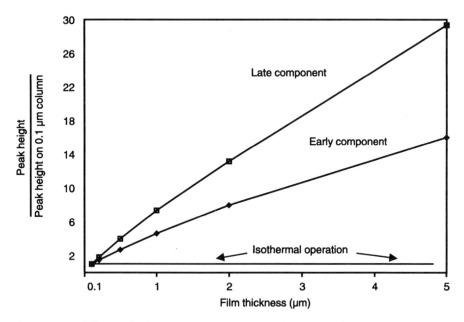

Figure 8.3 Effect of film thickness on peak height under temperature programming conditions. Calculated for 25 m x 0.25 mm columns programmed 50 to 300 °C at 2 °C min^{-1}. The sample loading is assumed to be increased in the same proportion to the film thickness

programming rates are constant and that the sample loading increases in direct proportion to the film thickness. The peak heights are calculated relative to those obtained from the 0.1 μm column. The peak enhancement increases clearly as the film thickness increases, and is greater for later eluting compounds than for the earlier ones.

Mean linear carrier gas velocity, $\bar{u}$

According to (8.20) the response of instruments utilizing mass flow sensitive detectors increases in proportion to $\bar{u}$. Fortunately, helium or hydrogen are preferred carrier gases for open tubular columns, and these are operated at faster velocities than nitrogen. Also, as plate number only declines slowly at velocities above the optimum, it is usually safe to use carrier velocities at up to twice the optimum. For concentration detectors, $\bar{u}$ does not appear in (8.18) and so any changes in flow rate will only affect the peak height slightly as a result of its effect on the column plate number.

Column length, L

R_D is proportional to $\sqrt{N}$ and inversely proportional to L, both for mass flow and concentration detectors, and so the nett effect is that detectability becomes proportional to $1/\sqrt{L}$. However, this would tend to be compensated by an equivalent increase in sample capacity so the overall effect on response would be entirely a result of any improvement in resolution.

8.3.4 General Recommendations for Trace Analysis

The following aspects should be noted in order to maximize the instrument capability for trace analysis.

1. Ensure that the column is correctly installed, thoroughly conditioned and operated with helium or hydrogen carrier gas at 1–2 times the optimum gas velocity. Ensure that the column plate number is satisfactory for the type of column.
2. Choose a medium film thickness of 0.5 μm or thicker in the case of volatiles.
3. Optimize the column conditions for the proposed analysis, ensuring that there is sufficient resolution between all components in the test mixtures. Use temperature programming if necessary. Change the stationary phase, if necessary, to increase the separation factor between the major and minor compounds. Use a thicker film in case of difficulty, but only where temperature programming is employed.
4. Ensure that the detector is operating under optimum conditions according to the manufacturer's manual.

5. Determine the minimum detectabilities for the components to be determined and assess the feasibility of carrying out the analysis without preconcentration steps.
6. Use either splitless or preferably on-column injection for concentrations of less than 10 ppm. Direct injection can be used for wide bore columns.

8.4 THEORY AND PRACTICE OF PROGRAMMED TEMPERATURE OPERATION (PTGC)

In Chapter 6 we saw how programmed temperature operation (PTGC) is used in splitless and on-column injection to assist the solvent effect in focusing the injection zone.

The main reason for the use of programming, however, is to compensate for the **chromatographic elution problem**. This is a consequence of (4.35) and (4.36) which show an exponential relationship between retention time and molecular weight for components of similar chemical type. This means that wide boiling range samples will give unacceptably long analysis times if they are run under isothermal conditions. This problem is overcome in practice by increasing the column temperature during the run, and most modern equipment can apply a range of appropriate gradients. **Linear temperature programming (LPTGC)**, in which the rate of temperature increase is kept constant, is the most commonly used operational mode. PTGC is normally characterized by an **initial temperature**, a rate of temperature increase or **programming rate**, usually specified in degrees per minute, and a **final temperature**. Both the initial and final temperatures can be maintained for isothermal periods according to the analytical requirements. While the theory of PTGC is not particularly complex, it is quite difficult to apply it in practice due to the involvement of exponential integrals, and the fact that the mean gas velocity varies with time because of viscosity and thermal expansion changes. Establishing suitable programming conditions for any particular analytical application is therefore usually carried out empirically, although the theory does provide some useful ground rules.

Alternatively, there are several commercial software packages available which can be used on personal computer systems and which remove most of the guesswork. The 'Drylab GC' commercial package, for instance, has been described in several papers [4,5] and is particularly useful for method development as it only requires the input of practical data from two PTGC runs to enable the prediction of retention characteristics, resolution and separation factors for other columns and programming conditions. Thus optimum conditions can be quickly derived in a fraction of the time that would be necessary using empirical methods. Figure 8.4a and b shows two such runs for a pesticide sample at 5 and 25 $°C\,min^{-1}$ ramp rate, on an open tubular

ANALYSIS AND OPTIMIZATION 253

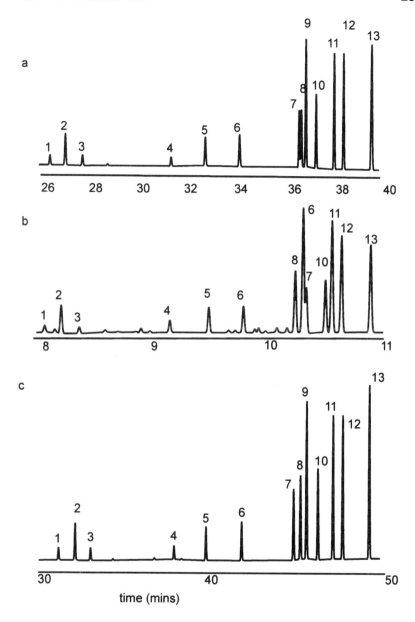

Figure 8.4 Experimental chromatograms from the analysis of standard pesticides sample using programming rates of (a) 5 °C min^{-1}; (b) 25 °C min^{-1} and (c) the optimum value of 3.9 °C min^{-1}. In all three cases the initial temperature was 35 °C and the final temperature was 330 °C. [Reproduced by permission of Advanstar Communications, from G. B. Abbay *et al. LC-GC International*, **4**, 28 (1991)]

column programmed from 35 to 330 °C. The programme computes two thermodynamic constants from these runs which then allows predictions to be made for other conditions. A **resolution map**, as shown in Figure 8.5 is a plot of the resolution between the most critical pair of components versus programme rate and an optimum rate is then easily derived from the plot. In this case a programme rate of 3.9 °C min^{-1} would appear to give a baseline separation of all components, and this is verified by the chromatogram shown in Figure 8.4c.

8.4.1 Linear PTGC Theory

The theory of LPTGC has been considered by a number of workers including Giddings, Harris and Habgood, Rowan, and Grant and Hollis [6–9].

Consider a single solute applied to the inlet of a column at an initial temperature of T_0 K, and then a linear programme is applied at r °C min^{-1} until the solute elutes. As the temperature increases so the solute zone accelerates due to the reduction in the retention factor k.

Suppose the zone centre travels dx cm in dt minutes, then:

$$dt = \frac{dx}{u_x}(1 + k_T) \tag{8.24}$$

$$= \frac{dx}{u_x}\left(1 + \frac{K_T}{\beta}\right) \tag{8.25}$$

where k_T = the retention factor of the solute at column temperature T and u_x = the linear gas velocity at distance x along the column and K_T = the distribution coefficient at temperature T.

K_T is related to the column temperature by the thermodynamic relationship given in (4.35), namely,

$$\ln(K_T) = \frac{H_s}{RT} + C \tag{8.26}$$

or

$$K_T = a\,e^{\frac{H_s}{RT}} \tag{8.27}$$

where a is a thermodynamic constant.
Therefore

$$dt = \frac{dx}{u_x}\left(1 + \frac{a}{\beta}e^{\frac{H_s}{RT}}\right) \tag{8.28}$$

ANALYSIS AND OPTIMIZATION

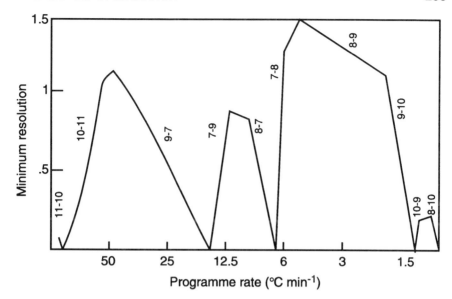

Figure 8.5 Resolution map of the pesticide sample of Figure 8.4. [Reproduced by permission of Advanstar Communications, from G. B. Abbay et al. LC-GC International, **4**, 28 (1991)]

Under LPTGC conditions, the column temperature after time t is given by

$$T = T_0 + rt \tag{8.29}$$

namely,

$$dT = r\, dt \tag{8.30}$$

which on substitution into (8.28) gives,

$$\frac{dT}{r} = \frac{dx}{u_x}\left(1 + \frac{a}{\beta}\, e^{\frac{H_s}{RT}}\right) \tag{8.31}$$

i.e.,

$$\int_{T_0}^{T_R} \left(\frac{dT}{1 + \frac{a}{\beta}\, e^{\frac{H_s}{RT}}}\right) = r\int_0^L \frac{dx}{u_x} \tag{8.32}$$

Where T_R = retention temperature of the solute (K).

The values of H_s and a are thermodynamic constants which can be evaluated from isothermal runs at two temperatures by using the isothermal relationship

$$t_R = t_0(1 + k) \tag{8.33}$$

where

$$k = \frac{a}{\beta} e^{\frac{H_s}{RT}} \tag{8.34}$$

The integral expressed in (8.32) does not have an exact solution, although it can be solved quite easily for practical purposes by computer integration. Thus, retention temperatures, separation factors and consequent resolutions can be predicted using a simple basic programme on a PC without the need for purchasing expensive software. The effect of variables such as gas velocity, starting temperatures, programming rates and column parameters can also be determined.

To produce general guidelines for LPTGC we need to make several simplifications and assumptions, as follows.

(a) The solute travels a negligible distance along the column during the holdup time and so the holdup time can be regarded as negligible. This will be the case for most compounds provided that the initial temperature T_O is sufficiently low.
(b) The mean carrier gas velocity $\bar{u}$ remains constant during the programme. This is a more serious assumption than (a) since certainly the velocity decreases during the run, although with pressure control there will be some compensation for the opposing effect due to the thermal expansion of the gas.

An approximate solution to (8.32) is therefore:

$$\frac{R}{H_s} \left(\frac{T_R^2}{e^{\frac{H_s}{RT_R}}} - \frac{T_0^2}{e^{\frac{H_s}{RT_0}}} \right) \cong \frac{raL}{\beta \bar{u}} \tag{8.35}$$

According to assumption (b) the second exponential expression is negligible except for very fast eluting components, i.e.,

$$\frac{RT_R^2}{H_s e^{\frac{H_s}{RT_R}}} \cong \frac{raL}{\beta \bar{u}} \tag{8.36}$$

This equation can be transposed into the following form:

$$T_R \cong \frac{H_s}{R\left(\ln \dfrac{RT_R^2}{H_s} - \ln\left(\dfrac{arL}{\beta \bar{u}}\right)\right)} \qquad (8.37)$$

The molar heat of solution H_s is approximately proportional to the molecular weight of the solute for similar compounds and so we would expect the retention temperature T_R to increase with H_s. In fact, variation of the term $\ln(RT_R^2/H_s)$ is virtually negligible compared with the change in retention temperature itself, and so we can state (8.37) in the following form:

$$T_R \cong \frac{H_s}{R\left(\text{const} - \ln\left(\dfrac{arL}{\beta \bar{u}}\right)\right)} \qquad (8.38)$$

It is **not** recommended that (8.38) should be used for any serious prediction of retention characteristics because it is an essentially approximate relationship which will show enormous variations for T_R for even very small changes in the denominator terms. It does give rise to some basic operational guidelines, however, and these are summarized as follows

Effect of boiling point on retention temperature

Retention temperature increases linearly with increasing molecular weight for chemically similar compounds. According to the well-known Trouton rule, the molar heat of vaporization at the boiling point is related to the absolute boiling point of the component by the vaporization entropy which has a numerical value of 20–23 for many compounds. On non-polar stationary phases, the heat of vaporization is similar to the heat of solution and so a plot of retention temperature versus boiling point will also be linear provided the components are chemically similar. On polar stationary phases where there may be considerable interaction between the solutes and the phase, the solution entropy may vary from this range of values, but for similar compounds it would still be fairly constant and so a linear relationship would still apply.

Figure 8.6 shows several practical plots obtained under a variety of conditions, to show the validity of this linear relationship. The linearity of plots of retention temperature versus boiling point forms the basis of the method of **simulated distillation**, which is discussed in Section 8.4.2. Note that the relationship becomes curved for the early components, or in the case of the acid esters the plot is curved throughout the range. Curvature will be most pronounced whenever the components start to migrate significantly along the column during the holdup time as this decreases the validity of (8.38), which assumes insignificant migration during this period.

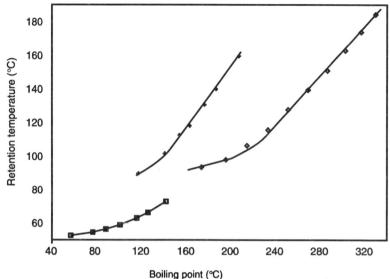

□ Acetic acid esters,10m x 53mm(5) PEG, Prog 50 to 100 °C at 5 °C min^{-1}
♦ Free fatty acids C2–C7, 25m x 32mm (5) poly(methylsiloxane), Prog 70 to 300 °C at 10 °C min^{-1}
+ n-Alkanes C10–C19, 25m x 32mm(0.12)poly(methylsiloxane), Prog 80 to 320 °C at 10 °C min^{-1}

Figure 8.6 Relationship between retention temperature and boiling point for three different chemical classes

Predicted effects of variables

The effects of programme rate, column length, column diameter and stationary phase film thickness can all be evaluated from the denominator term $\ln(arL/bu)$. Thus, increasing the programme rate, column length or film thickness all **increase** the retention temperature, while increasing the column diameter or the gas velocity all **decrease** the retention temperature. It is interesting to note that provided this logarithmic term is maintained constant then there should be little effect on the retention temperature. Thus, if the film thickness is increased and the programme rate decreased in inverse proportion, then the retention temperature will remain approximately constant. The retention time would not remain the same, of course, because of the different programme rate. For example, if the film thickness were increased from 1–2 μm and the programme rate kept at the same value then both the retention temperature and retention time would increase. If, however, the programme rate is halved for the second column, the retention temperature would remain the same but the analysis time would be doubled.

Figure 8.7 shows a comparison of the experimental effect of programme rate on the retention temperatures of several hydrocarbons with values calculated from the above theory. The column was a 25 m x 0.25 mm open

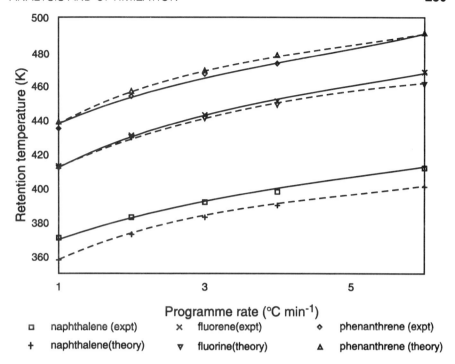

Figure 8.7 Relationship between retention temperature and programme rate for hydrocarbons: comparison with theory

tubular column coated with a 1 μm film of poly(methylsiloxane) stationary phase. The starting temperature was 80 °C. There is generally good agreement between the theoretical and experimental values.

For practical optimization, a trial run should first be carried out using empirical conditions based on the boiling range of the sample. A starting temperature of 40–50 °C below the boiling point of the lowest boiling component is suggested, at a nominal programme rate of 5 °C min^{-1}. The final temperature should be lower than the MAOT value for the column. Refer to Table 8.2 for subsequent adjustments based on the results of the pilot run.

8.4.2 Simulated Distillation

As the name implies, simulated distillation is the use of gas chromatography for boiling point separations of complex samples to characterize component distributions in terms of their boiling points. Ideally, the technique should produce similar results to those that would be obtained by high efficiency fractional distillation, but considerably faster. The technique has found widespread application in the petrochemical industry where many types of oil may need to be characterized rapidly, perhaps according to a product

Table 8.2 Optimization of programming conditions

Problem	Action
Early components too rapidly eluted with insufficient resolution	Use an initial isothermal period and/or lower the initial temperature If this is already very low, then use thicker film of stationary phase
Retention times of early components very long	Raise the starting temperature
Last components elute after the temperature programme is completed	Raise the final temperature (provided this does not exceed the MAOT) Decrease the programme rate
Final components elute well before the final temperature	Lower the final temperature Increase the programme rate if resolution OK
Components all elute within the programme but overall resolution poor	Decrease the programme rate and lower the final temperature Use another stationary phase
Excessive baseline rise during programme	Condition column thoroughly Ensure temperature does not exceed the MAOT value Rinse column or replace it

specification. Automatic commercial equipment is available for this purpose.

Simulated distillation is based on the correlation between retention time and boiling point which, as discussed earlier, is approximately linear under linear programming conditions. Thus, columns chosen for this technique are usually non-polar, and capable of operating over a wide temperature range. Individual component separation is not required for this technique and so the column plate number does not have to be very high. Consequently short wide bore stainless steel columns, coated with a high stability polysiloxane phase are the preferred choice. A typical example of such an application is shown in Figure 8.8 to characterize the boiling range of a motor oil. The sample was spiked with paraffins in the range C15–C24.

Programming column conditions are first set to elute the samples in a reasonably short period of time, and to give an approximately linear correlation between the boiling point and retention time. This relationship is established by calibration with a sample of known composition covering the same boiling range as the samples. The required boiling ranges for the simulated distillation runs can then be equated to retention time intervals.

On running the sample the chromatogram is integrated continuously to give cumulative peak areas at these retention time intervals, and these area are converted to percentages of the total oil from response factors determined with a known boiling range oil.

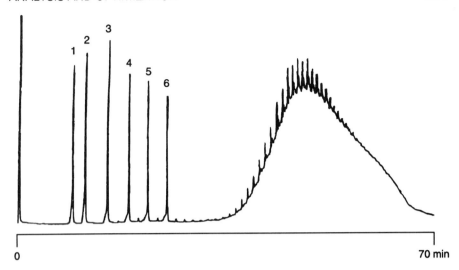

Figure 8.8 Simulated distillation chromatogram for motor oil spiked with hydrocarbons. Column: 10 m × 0.5 mm 'Ultimetal' stainless steel coated with high stability polysiloxane phase. Temperature: 40 to 400 °C at 5 °C min^{-1}; carrier gas: helium at 3.5 ml min^{-1}. Spiked compounds: 1 = pentadecane; 2 = hexadecane; 3 = octadecane; 4 = eicosane; 5 = docosane; 6 = tetracosane. (Reproduced by permission of Chrompack International BV)

8.5 PRACTICAL OPTIMIZATION AND TROUBLESHOOTING

The theoretical principles of optimization have been discussed in Chapters 2 and 4 where the effects of column parameters and operating conditions such as choice of carrier gas and carrier gas velocity were covered. This section is intended for more practically orientated users, or those who require the use of capillary GC as a tool without the necessity of more detailed knowledge.

The section is organised as a problem and solution dialogue with some illustration of the situations presented. Users should follow the recommendations in strict order. Only the more common troubleshooting problems are dealt with here and the reader is referred to the series of articles by J. Hinshaw which appear regularly in *LC-GC International* on this topic. A recent book by Rood also deals in depth with troubleshooting problems [10]. Modern capillary instrumentation is relatively reliable and most of the practical problems are due to the operator's choice of conditions, lack of attention to details such as checking for leaks, blockages, etc., ensuring that septa are replaced regularly, cleaning injectors, and maintaining detectors in a good condition. **Prevention of problems by a high standard of 'housekeeping' is always more effective than dealing with problems after they occur.**

Problem 1
Peaks appear to be broad, perhaps with tailing and resolution is poor

Make a visual check of the operating conditions, particularly the temperatures of the injector, column and detector. Low injector temperatures are a particular cause of this problem.

Ensure that the system is not contaminated, namely, contamination of the injection liner by sample decomposition products, septum fragments, etc. The injection septum should be replaced frequently.

Ensure that carrier gas is flowing at the appropriate velocity by injecting 0.1 ml methane (natural gas) from a gas tight syringe.

Calculate the linear gas velocity by dividing the column length by the retention time for the methane peak. Ensure that it agrees with the recommended data given in Table 4.2.

Check the appearance of the methane peak which should be sharp. If the peak shows any tailing then either the column installation is faulty or the split flow rate is too low. Check for leaks at the connections and the positioning of the column inside the detector and injector.

Measure the column plate number from equation (1.9). This should be roughly equal to the length of the column divided by its diameter in the same units. If the plate number is very low and the column has been in use for some time then replace it. (Chemically bonded phases can usually be rinsed. This will often regenerate the performance of columns.)

Ensure that the sample volume is not excessive. Injection systems have a limited capacity and any attempt to use an excessive volume of sample will cause back diffusion of the vapour, peak tailing, particularly of the solvent, and contamination of the equipment giving rise to memory effects and ghost peaks.

Problem 2
No apparent peaks after sample injection
Check in order,

- Is the flame detector ignited?
- All relevant instrument switches and settings
- Possible blockage of the syringe
- The syringe seal
- Carrier gas flow? Inject methane to see if peak appears; if no peak, check for blockages and leaks in the system.

Problem 3
Peaks start to tail after prolonged column usage

Check to see if this is due to the column, or to the injector/detector. Examine the injector liner and ensure it is clean.

ANALYSIS AND OPTIMIZATION

Usually, tailing peaks are a sign of column contamination or degradation. It may be possible to regenerate the column by rinsing it if the phase is bonded, otherwise replace the column.

Problem 4
Peaks tail, even for new column

Assuming correct column installation, etc., in accordance with the previous section, this is an indication that the tailing is due to interaction with the interior surface of the column. There are two possible remedies.

- Use a thicker film of stationary phase.
- Use a more polar stationary phase.

If a PLOT column is in use then peak tailing can result from sample overloading. In this case reduce the sample size.

Problem 5
Fronting peaks, i.e., the leading edge is more diffuse than the trailing edge

This is almost certainly due to column overloading. Reduce the sample size either by increasing the split, if used, or by diluting the sample further.

Problem 6
Excessively noisy baseline

There are numerous possible cause of this problem, but the most common reason is excessive stationary phase bleed. This could arise from the following causes.

- Column temperature is too high. Lower the temperature.
- Contamination of the column. Try running the column at an elevated temperature close to the MAOT value overnight.
- The stationary phase is decomposing due to the presence of oxygen in the carrier gas. The column may need to be replaced.
- Ensure that an efficient oxygen filter is being used in the carrier gas line and check for possible leaks.
- Particles may be dislodging from PLOT columns. These are usually supplied with particle traps. Particles will give random spikes on the baseline.

Other causes of baseline instability are as follows.

- Injector contamination. Ensure that the liner is clean and free from septum fragments, etc.
- Detector contamination. This can be caused by stationary phase deposits. Is the detector temperature high enough? Other forms of detector contamination arise from fragments of column, deposits of silica from polysiloxane

phases, carbon deposits, particularly if the solvent is hydrocarbon. In the latter cases, the detector should be cleaned and serviced.
- Electronic or electrical faults. Faulty connections and any vibration or movement of the amplifier detector connections may give baseline disturbances.

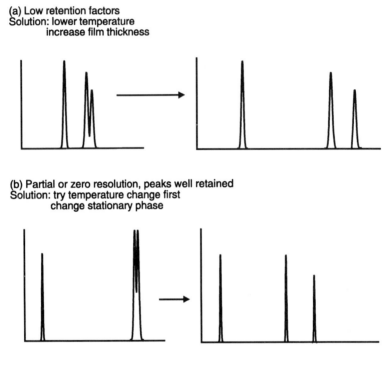

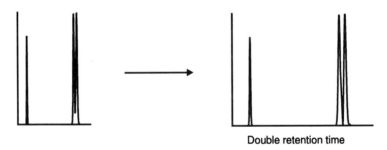

Figure 8.9 Some basic optimization situations

ANALYSIS AND OPTIMIZATION

Problem 7
Drifting baselines

This usually is due to column bleed or programmed temperature operation. Ensure that the column is thoroughly conditioned. Contaminated detectors, injectors, leaks and blockages can also cause baseline drift.

Problem 8
Good peak shape and satisfactory plate number but the resolution is poor

This is more of an optimization problem and some of the simplest situations are depicted in Figure 8.9.

8.6 GENERAL CONCLUSIONS

This chapter has dealt with some of the more basic features of practical capillary GC, particularly from the viewpoint of the user who requires a quick result without the necessity of a more detailed study.

With regard to quantitative analysis, the internal standard method should be used for liquid samples for maximum accuracy and precision. For the analysis of gases the simple external standard method can be used with considerable accuracy. In any case, the user should always be aware of the large number of variables which can affect both chromatographic properties and analytical accuracy. All the components of interest should give symmetrical peaks and be resolved to baseline if possible, i.e., with a resolution of at least 1.5.

The injection procedure can have a major effect on analytical error and merits particular care and attention. The injection system must be kept in a clean condition, and septa should be changed frequently. Other basic precautions are the correct installation of the column and a check to ensure the absence of any leaks. The microlitre syringe should also be checked frequently to ensure that is in good leak-proof condition.

For trace analytical applications, preliminary tests should be carried out to determine the minimum detectabilities of the components of interest, and consequently to decide on any sample preparation, and injection volumes. The detection of low concentrations will be more successful with a stable system and detector operated under optimum conditions rather than simply to overload the injector with more and more sample. Better detectability results with thicker film columns run under programmed temperature conditions.

Use programmed temperature conditions for samples having a wide boiling range, i.e., greater than 80–100 °C. Adjust the conditions following a pilot run according to the recommendations given in Table 8.2.

REFERENCES

1. Jeffery, P. G. and Kipping, P.J. (1972) *Gas Analysis by Gas Chromatography*, 2nd edn, Pergamon Press, Oxford.
2. Cowper, C. J. and DeRose, A. J. (1983) *The Analysis of Gases by Chromatography*, Pergamon Press, Oxford.
3. Barrett, R. S. (1981) *Analyst (London)*, **106**, 817.
4. Bautz, D. E., Dolan, J. W. and Snyder, L. R. (1991) *J. Chromatogr.*, **541**, 1.
5. Abbay, G. N., Barry, E. F., Leepipatpiboon, S., Roman, M. C., Siergiej, W., Snyder, L. R. and Winniford, W. L. (1991) *LC–GC International*, **4** (3), 28.
6. Giddings, J. C. (1962) *Anal. Chem.*, **34**, 722.
7. Harris, W. E. and Habgood, H. W. (1966) *Programmed Temperature Gas Chromatography*, Wiley, New York.
8. Rowan, R. (1967) *Anal. Chem.*, **39**, 1158.
9. Grant, D. W. and Hollis, M. G. (1978) *J. Chromatogr.*, **158**, 3.
10. Rood, D. (1991) *A Practical Guide to the Care, Maintenance, and Troubleshooting of Capillary Gas Chromatographic Systems*, Chromatographic Methods Series, Huethig, Heidelburg.

9
Multidimensional Capillary GC and Column Switching

9.1 INTRODUCTION

Multidimensional gas chromatography (MDGC) can be defined as any GC separation process carried out with an integrated system of two or more columns. Packed column MDGC systems have been in use for many years to perform complex separations in the petrochemical industry [1–3], particularly for the analysis of wide boiling range oils containing various classes of compound. Attempts to separate mixtures of this type on a single column are rarely completely successful and even the use of several different independently operated columns is unlikely to improve the situation because of peak shuffling effects. Multidimensional systems, however, in which the columns are combined by column switching, can optimize the separation capabilities of the different columns. Typical situations which can benefit from the use of multidimensional GC are as follows.

- The separation of components which cannot all be separated by a single column.
- Selective separation of regions selected from a precolumn separation. This process is known as **heart-cutting**.
- Separation of the most volatile part of a wide boiling range sample. The residue sample is then removed by reversing the direction of carrier gas flow through the precolumn, a process known as **back-flushing**. This is also carried out on single columns as a way of venting high boilers.

Some multidimensional systems are completely automatic and designed for specific applications, as in analysers for particular products. Others are constructed for more general applications of the above three categories. The cost of this type of commercial equipment, however, should not be a deterrent to the potential user as it is possible to adapt any laboratory capillary chromatograph to multidimensional operation by using commercially available components, such as appropriate valving, etc.

There are two major problems to overcome in the design of multidimensional systems, i.e., the problem of the compressible nature of the carrier gas, and the need to keep extra-column volumes to negligible proportions. The first of these problems requires that any column switching operations should not cause significant changes in the pressures at different points in the system. One exception to this rule is in the event of back-flushing, where the flow rate through a preliminary column is reversed to discard sample residue, but this would only be applied after the main separation function of the column has been completed. The necessary pressure stability is accomplished by using empty capillary tubes with similar dimensions to those of the columns in the system, to provide a balancing resistance when the columns are switched in and out of the sample flow path.

The second problem of avoiding significant extra-volume effects requires the use of either precision switching valves specifically manufactured for capillary GC, or switching mechanisms that do not involve any direct contact with the sample. Commercial valves are manufactured for capillary-sized columns and will fulfil these requirements, but an alternative method is based on the use of **pressure balancing systems**. Deans first described a method of heart-cutting using packed columns based on this principle [1,4] and this can be extended easily to open tubular columns. Packed columns are still used in several commercial analyser systems, but there is a trend to replace these with open tubular columns incorporating the Deans' switch to gain the benefits of faster analysis and better separation.

There are, of course, numerous possibilities for MDGC ranging from simple two column arrangements connected in sequence without switching, to complex systems involving several columns and precisely timed switching facilities. However, MDGC is not a widely utilized technique in general because the enormous resolving power of the modern column is usually adequate for the majority of applications. Nevertheless, there are practical situations which can benefit from the use of combined columns and some of these will be discussed.

9.2 SERIES CONNECTED COLUMNS WITHOUT SWITCHING

Two columns connected together in series and without flow switching is the simplest possible arrangement for a multidimensional system, and its main function would be to optimize the separation of complex mixtures by combining the selective properties of two different stationary phases. Nevertheless a two-dimensional system is also set up whenever a broken column is repaired, or if two different columns are joined together to increase the overall column length. Provided the two lengths have the same diameter, are coated with the same stationary phase and have the same film thickness then the column plate number will be the sum of the individual plate numbers **provided that the plate height is the same in both columns.**

There is, however, an inherent danger in connecting open tubular columns together in this way. For instance, if the two columns have different plate heights then the final plate number will not simply be the sum of the two values for the separate columns.

Consider a number of columns A, B, C, etc., connected in series. The summation of variance law states:

$$\sigma_{tot}^2 = \sigma_A^2 + \sigma_B^2 + \sigma_C^2 + \text{etc.} \tag{9.1}$$

where σ_{tot} = the standard deviation in time units of a peak eluting from the last column; σ_A, σ_B, etc., are the standard deviations of the peaks eluting, respectively, from columns A, B, etc.

Confining our attention to two columns, the overall plate number is given by,

$$N_{tot} = \frac{t_R^2(\text{tot})}{\sigma_{tot}^2} \tag{9.2}$$

$$= \frac{t_R^2(\text{tot})}{(\sigma_A^2 + \sigma_B^2)} \tag{9.3}$$

The standard deviations of eluting peaks can be derived from the standard deviation of the zone when it reaches the end of the column, and its speed of elution. This is $\sqrt{N}$ plates according to the plate theory, equivalent to a distance of $H \times \sqrt{N}$ cm. As the zone is migrating at a speed of $\bar{u}/(1+k)$ cm s^{-1}, then,

$$\sigma_A = \frac{L_A(1+k_A)}{\bar{u}_A \sqrt{N_A}} = \frac{t_R(A)}{\sqrt{N_A}} \tag{9.4}$$

and

$$\sigma_B = \frac{L_B(1+k_B)}{\bar{u}_B \sqrt{N_B}} = \frac{t_R(B)}{\sqrt{N_B}} \tag{9.5}$$

namely,

$$N_{tot} = \frac{t_R^2(\text{tot})}{\frac{t_R^2(A)}{N_A} + \frac{t_R^2(B)}{N_B}} \tag{9.6}$$

Equation (9.6) shows that if, for example, column A is a poor column with a lower plate number than column B then the total plate number will be lower than the sum of the two plate numbers, particularly if the retention characteristics on column A are much longer than on B. In relation to MDGC, it is clearly inadvisable for column A to be wider in diameter than B unless each zone is refocused between the two columns. This can usually be performed by the use of either cryogenic focusing or the use of a small bed of absorbent, as discussed in Chapter 7.

9.2.1 Selectivity Tuning with Series Connected Columns

Direct serial coupling of two columns as a way of optimizing difficult separations is potentially a very powerful technique which is not in fact widely practised. This is surprising in view of the accuracy with which commercial open tubular columns can now be produced, and the ease with which such columns can be joined together.

The total retention time of a component passing through two columns connected serially is given by,

$$t_R(\text{tot}) = t_R(A) + t_R(B) \qquad (9.7)$$

$$= \frac{L_A(1 + k_A)}{\bar{u}_A} + \frac{L_B(1 + k_B)}{\bar{u}_B} \qquad (9.8)$$

L_A and L_B are the respective lengths of columns A and B; $\bar{u}_A$ and $\bar{u}_B$ are the mean gas velocities through the two columns. These mean velocities will not be the same for the two columns because of the different pressure drops down each column, and the compressibility effects. However, calculation shows that there would be a maximum difference in average gas velocity of less than 0.5% for open tubular columns of normal length and diameter. Therefore, we can assume for practical purposes that $\bar{u}_A = \bar{u}_B$.

One practical use of selectivity tuning by column coupling is to develop new phases with optimum separation characteristics toward specific types of

---

Figure 9.1 Separation of EPA mixture 624. Columns: 25 m × 0.32 mm (1.2 μm); temperature 35 to 250 °C at 4 °C min^{-1}. (a) Stationary phase: CPsil8CB [poly(5% phenylmethylsiloxane)]. Unresolved components: benzene (10) and carbon tetrachloride (11); 1,2-dichloropropane (12) and trichloroethylene (14); trans-1,3-dichloropropylene (17) and 1,1,2-trichloroethane (18). (b) Stationary phase CPsi19CB [poly(7% phenyl 7% cyanopropyl methylsiloxane)]. Unresolved components: methylene chloride (4) and trans-1,2-dichloroethylene (5); 1,1,1-trichloroethane (9) and carbon tetrachloride (11); 2-chloroethylvinylether (15) and cis-1,3-dichloropropylene (16); dichlorobenzene (22) and ethylbenzene (23). (Data reproduced by permission of Chrompack International BV)

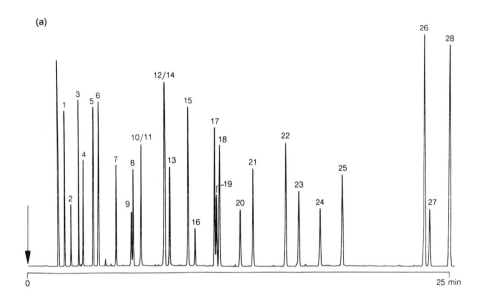
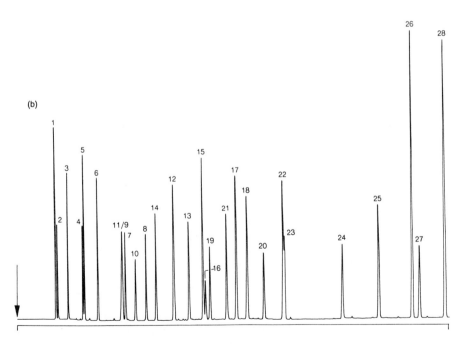

complex samples. A typical example is the separation of volatile halogenated hydrocarbons, as specified by EPA 624 for environmental samples.

Figure 9.1 shows the separation of a test mixture containing the specified compounds on (a) a 5% phenyl 95% dimethylsiloxane and (b) a polysiloxane phase containing 7% phenyl and 7% cyanopropyl. Both columns were 25 m in length, with a diameter of 0.32 mm, and film thickness of 1.2 μm.

Although most components are resolved to baseline, there are still some partly or unresolved groups on both columns and these are identified in Figure 9.1. Since the unresolved groups differ on the two columns, there is a good chance that a column with intermediate selectivity could resolve all of the 28 compounds.

Equation (9.8) shows that the total retention time for each component varies linearly with the fractional lengths of the two columns, and so an optimum separation should be possible by connecting the appropriate lengths together. These lengths could be found from a graph of total retention time versus fractional length of the column. However, with so many compounds this produces a confusing network of lines and to find optimum regions would be very difficult from visual examination.

A considerably better method of finding the optimum lengths is to compute the resolution of all possible pairs of components over the entire range of the respective fractional lengths, and to plot a **minimum resolution map**. This method is similar to that discussed previously in Chapter 8 in relation to temperature programming optimization and can be accomplished easily by a basic programme on a personal computer.

Figure 9.2 shows the resulting plot for the EPA 624 mixture of halogenated hydrocarbons.

This diagram indicates that there are several regions which should produce a baseline resolution of all the components but a fractional length of between 0.4–0.5 of column B covers the widest range and this was chosen for further investigation. Figure 9.3 shows (a) a computer simulation of the separation expected for a combined column comprising 47% column A and 53% column B, and (b) an experimental run in which the stationary phase was mixed in the same ratio.

The experimental separation agrees closely with that predicted from the resolution map except for the incomplete resolution of benzene and dichloroethane. This is because (a) programmed temperature operation was used in practice whereas the computer simulation assumed isothermal operation, and (b) the separation was performed on a mixed phase rather than on separate columns combined in the same proportion. Although the retention times appear to behave in an additive manner on the mixed phase in this case, this cannot be extended to polar phases which may show high degrees of interaction.

In this example, we have seen how the use of combined columns can help in the development of an optimum stationary phase to carry out difficult

MULTIDIMENSIONAL CAPILLARY GC AND COLUMN SWITCHING 273

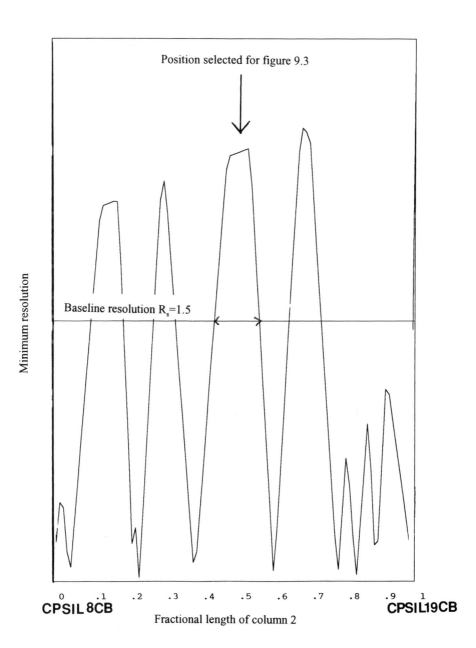

Figure 9.2 Minimum resolution map for bidimensional column system comprising of the columns used in Figure 9.1

274 CAPILLARY GAS CHROMATOGRAPHY

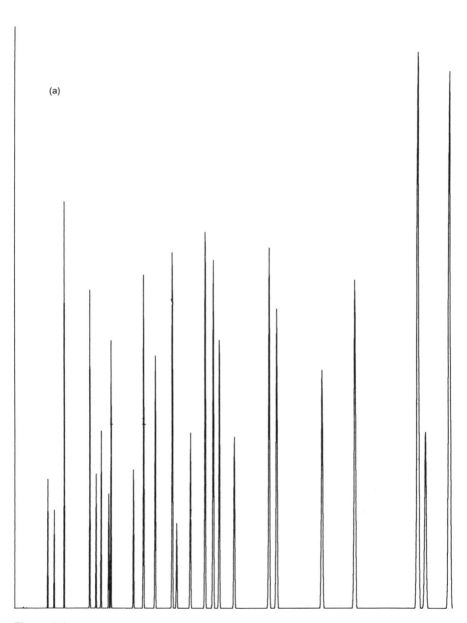

Figure 9.3

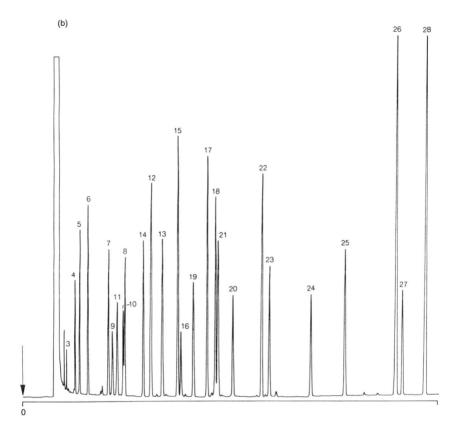

Figure 9.3 (a) Computer simulation of predicted chromatogram from the position shown in Figure 9.2 illustrating probable separation of all components. (b) Actual chromatogram obtained for phase mixed in the same proportions. (Reproduced by permission of Chrompack International BV.) Key: (1) chloroethane (not shown), (2) trichlorofluoromethane (not shown), (3) 1,1-dichloroethylene, (4) methylene chloride, (5) *trans*-1,2-dichloroethylene, (6) 1,1-dichloroethane, (7) chloroform, (8) 1,2-dichloroethane, (9) 1,1,1-trichloroethane, (10) benzene, (11) carbon tetrachloride, (12) 1,2-dichloropropane, (13) bromodichloromethane (14) trichloroethylene, (15) 2-chloroethylvinylether, (16) *cis*-1,3-dichloropropylene, (17) *trans*-1,3-dichloropropylene, (18) 1,1,2-trichloroethane, (19) toluene, (20) dibromochloromethane, (21) tetrachloroethylene, (22) chlorobenzene, (23) ethyl benzene, (24) bromoform, (25) 1,1,2,2-tetrachloroethane, (26) 1,3-dichlorobenzene, (27) 1,4-dichlorobenzene, (28) 1,2-dichlorobenzene

separations. A similar technique can be useful in general for separations of complex mixtures.

9.3 BACKFLUSHING TECHNIQUES

Samples often contain 'heavy ends' or residual components which are not required for analysis but need to be removed from the column to avoid contamination or prolonged analysis time. The simplest way to do this is to reverse the carrier gas flow rate through the column after the components of interest have eluted. Figure 9.4 shows a configuration using an eight port valve.

In (a), with the valve in the inject position, sample from the injector passes through the valve and then through the column, back though the valve and hence to the detector. A subsidiary flow of carrier gas is supplied at the same flow rate to the valve and hence to a restrictor to vent. When the components of interest have left the column, the valve is turned to the backflush position (b) which reverses the flow through the column to vent the residual sample. The flow rate through the detector is not affected as the restrictor has the same resistance to flow as the column.

A backflush system which performs the same function as Figure 9.4 but using pressure balancing for reversing the column flow is shown in Figure 9.5.

The system is set up initially with SV1 turned on and SV2 and SV3 turned off. The pressure from PC1 is set to give the required carrier gas velocity through the column and the natural midpoint pressure is noted on gauge G2. SV3 is then turned on and PC2 adjusted to give a **slightly** higher pressure than the natural midpoint pressure. For backflushing, SV1 is turned off to prevent

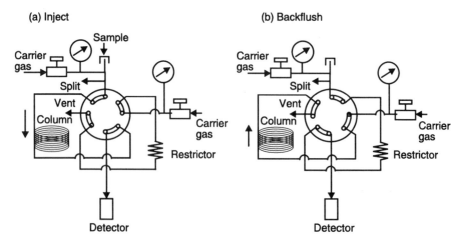

Figure 9.4 Backflush system using an eight-port switching valve

access of carrier gas to the column via PC1, and SV2 is turned on, opening the inlet side of the column to vent. Thus the pressure difference across the column is reversed due to the applied controlled pressure from PC2, and the flow through the restrictor to the detector is maintained at its previous level.

Although both of these diagrams show monodimensional systems they can be used equally as bidimensional systems for the removal of heavy ends. Thus

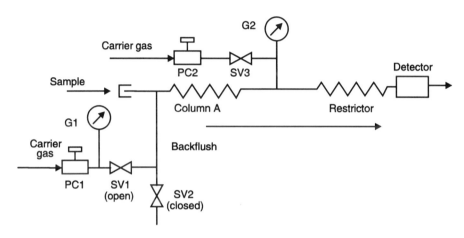

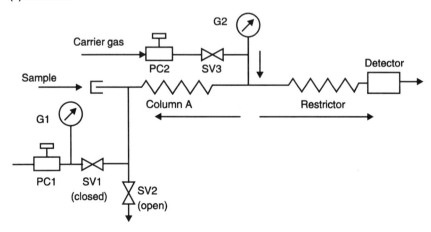

Figure 9.5 Backflush arrangement using the Deans' pressure balance system

the column shown could be a low efficiency precolumn such as a packed or wide bore thick film column, with a second high efficiency open tubular column connected immediately prior to the detector in Figure 9.4 or in place of the restrictor in Figure 9.5.

9.4 HEARTCUTTING TECHNIQUES

Heartcutting is a powerful technique that can be applied to a wide variety of samples. Some typical examples are as follows.

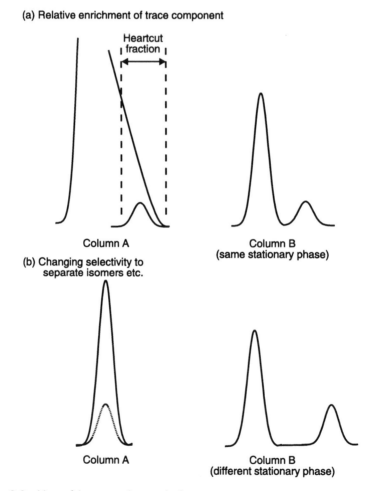

Figure 9.6 Use of heartcutting techniques

- Enrichment of trace components, particularly those in the tail of major components (Figure 9.6a).
- Analysis of co-eluting compounds from a preseparation column (Figure 9.6b).
- Elimination of large amounts of solvent, particularly if this causes problems in the second column detector. For instance, water would cause quenching effects in an electron capture detector.

A number of manufacturers supply multidimensional capillary systems, ranging from simple conversion kits to sophisticated instruments in which the separate columns can be accommodated in separate ovens enabling them to be operated under different temperatures conditions. The following options are usually available.

- 'Sampling' (heartcutting) of precolumn chromatograms at required positions and application of the fraction to a second high resolution column.
- Cryogenic or absorption focusing of the fraction eluting from the precolumn prior to its application to the second column by rapid vaporization.
- Backflushing of sample residue from the precolumn.

A simple practical system demonstrating the most commonly used configuration of two columns is shown in Figure 9.7. This system enables fractions to be selected from column A and passed to column B which would normally contain a different stationary phase capable of separating the components of the selected fractions. This particular arrangement does not permit any backflushing of column A and a more complex system is necessary for this. Commercial systems would, of course, include this facility as well as to allow separate injections into the two columns.

The reason for including the cold trap is evident from (9.6). Without cryogenic focusing the volume of carrier gas required to elute each fraction from column A would generate a long feed time to column B, thus reducing its efficiency. Assuming that we can tolerate a 20% loss of plate number then the following condition applies:

$$\frac{t_R^2(A)}{N_A} \leq \frac{0.2 t_R^2(B)}{N_B} \tag{9.9}$$

In most cases column A is a precolumn with a higher capacity than column B and so this condition will not be satisfied. Some form of focusing is then necessary, consisting in practice of either a cold trap, or a trap containing a suitable absorbent.

To operate this system with cryogenic focusing, the cold trap would be precooled to the requisite temperature using liquid nitrogen or CO_2 before switching the valve to the heartcut positions, when the required fraction will

(a) Injection-separation on column A

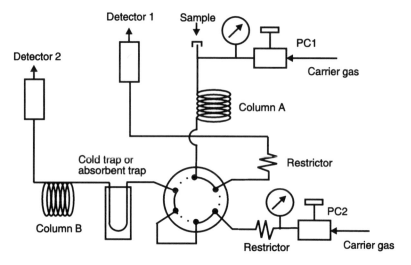

(b) Heart-cutting-fraction trapped and separated by column B

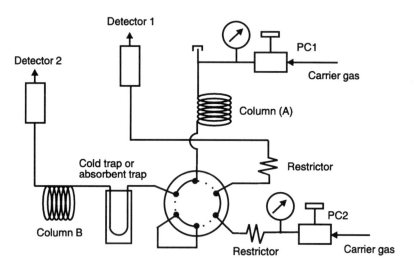

Figure 9.7 Basic multidimensional system employing two columns and switching valve

condense in the trap. At the end of the collection time the valve is reset to the inject position and the cold trap rapidly heated electrically to vaporize the sample and dispatch it as a sharp zone to column B.

Multidimensional systems based on the Deans' principle tend to be rather more popular, but they are restricted to isothermal operation. A typical arrangement is shown diagrammatically in Figure 9.8. Here, the flow of carrier gas from a preseparation column (A) can be directed either through the restrictor to the monitor detector 1 or through a cold trap to a second column (B), monitored by detector 2. The direction of flow is switched according to the direction of the applied pressure from the Deans' switch as activated by the three-way solenoid valve. Note that switching this valve does not affect the pressures at the inlets of the two columns and so there is no disturbance in the column flow rates or the detector response.

This type of configuration allows various types of column to be coupled, namely, packed to packed, packed to capillary and wide bore to normal bore open tubular. To determine trace components, for instance, column A could be a packed or wide bore column which would allow large samples to be analysed. Heartcutting is then carried out in the appropriate regions without the danger of overloading column B.

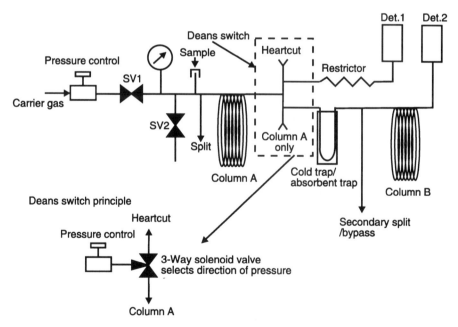

Figure 9.8 Diagram of a versatile multidimensional system using pressure balance switching

CAPILLARY GAS CHROMATOGRAPHY

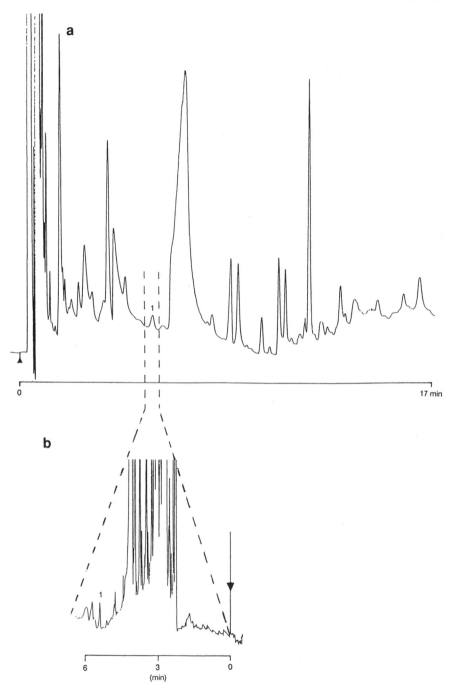

Two examples of the use of multidimensional capillary GC are shown in Figures 9.9 and 9.10. Figure 9.9 shows an application for the unambiguous determination of trace amounts of polyethylene glycol in wine. The preseparation on the 10 m wide bore precolumn enables the fraction containing ethylene glycol to be isolated and, after cryoscopic trapping, this is passed to a column containing a more polar stationary phase. In fact, only a minute part of the fraction is due to ethylene glycol. The other example shown in Figure 9.10 illustrates the use of the technique for the determination of trace levels of dioxane in a shampoo. The preseparation was carried out on a short wide bore column containing a non-polar stationary phase, and the heartcut further separated on a 25 m open tubular column coated with a highly polar phase, 1,2,3-tris(2-cyanoethoxy)propane (TCEP).

9.5 MULTICOLUMN ANALYSIS

Several types of complex sample in the petrochemical industry require complex column systems in order to obtain the necessary analytical information. Such systems are available as analysers from a variety of commercial sources for the analysis of natural gas, naphthas, refinery gases and products such as hydrocarbon oils.

Figure 9.11 shows a switching and column system commonly used for the analysis of natural gas. Although there are many possible types of natural gas analysis, the most important determinations are for nitrogen, carbon dioxide and lower hydrocarbons.

The arrangement shown uses two columns. Column A could either be coated with a non-polar phase, or it could be a porous polymer PLOT column. The sample is applied with the valve in the inject position where the two columns are connected in series. The permanent gases oxygen, nitrogen, helium (the diluent gas) and methane pass through column A quickly to column B, the molecular sieve 5A PLOT column, while heavier hydrocarbons are retained near the start of column A. The valve is then switched to the second position which removes the serial configuration, but continues with the separation of the two fractions.

Figure 9.9 Determination of diethylene glycol in wine using MD capillary GC and heartcutting. Precolumn conditions: (a) 10 m × 0.53 mm CPsil5CB (polymethylsiloxane) (5 μm); temperature: 100 (6 min) to 160 °C at 30 °C min^{-1}; splitless injection of 0.5 μl. Heartcut (b) is fraction 1 corresponding to the retention time for diethylene glycol. Analytical column conditions: 25 m × 0.25 mm cpwax 52CB (chemically bonded polyethylene glycol); temperature: 160 °C FID. Diethylene glycol content (peak 1) = 0.4 ppm. (Reproduced by permission of Chrompack International BV)

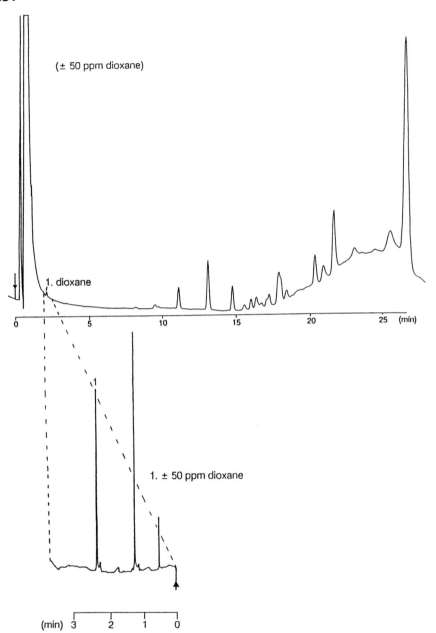

Figure 9.10 Determination of trace dioxane contamination of a shampoo. Precolumn (top) conditions: 10 m x 0.53 mm CPsil5CB (5 μm); temperature: 75 °C; splitless injection 0.5 μl of 1:1 methanol solution. Analytical column (bottom) conditions: 25 m x 0.25 mm TCEP [1,2,3-tris(2-cyanethoxy)propane], (0.4 μm); temperature: 75 °C. (Reproduced by permission of Chrompack International BV)

Position A for injection and group separation of methane + permanent gases from CO_2 and higher hydrocarbons

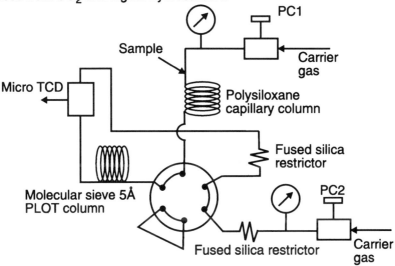

Position B for CO_2 and higher hydrocarbons

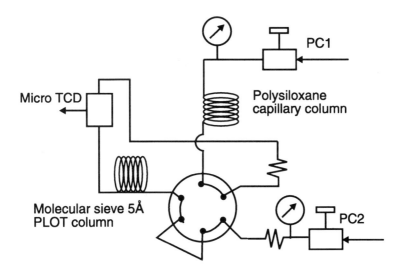

Figure 9.11 Schematic diagram of a system for natural gas analysis. [Reproduced from the *Journal of Chromatographic Science* by permission of Preston Publications, a division of Preston Industries, Inc., adapted from L. H. Henrich, *J. Chromatogr. Sci.*, **26**, 198 (1988)]

Figure 9.12 shows a typical chromatogram of natural gas. Peaks 1–5 are from column A and peaks 6–9 are from column B.

A commercial system for a complex analysis of refinery gas is shown in Figure 9.13 and consists of three columns, two of which are packed columns and one is a PLOT alumina column. This system performs an analysis of permanent gases, CO, CO_2 and C1–C8 saturated and unsaturated hydrocarbons by selecting the appropriate columns at the requisite times and presenting the

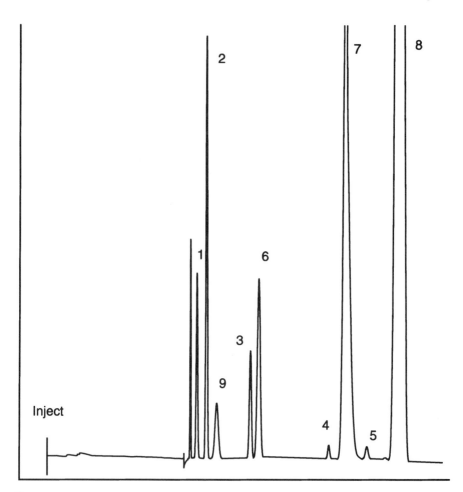

Figure 9.12 Example of the use of a two-column system for analysis of natural gas. Peaks 1 = CO_2; 2 = ethane; 3 = propane; 4 = isobutane; 5 = n-butane; 6 = oxygen; 7 = nitrogen; 8 = methane; 9 = helium. [Reproduced from the *Journal of Chromatographic Science* by permission of Preston Publications, a division of Preston Industries, Inc., adapted from L. H. Henrich, *J. Chromatogr. Sci.*, **26**, 198 (1988)]

MULTIDIMENSIONAL CAPILLARY GC AND COLUMN SWITCHING

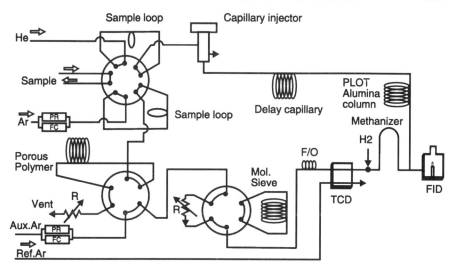

Figure 9.13 Diagram of a commercial system for refinery gas analysis. (Reproduced by permission of Varian Analytical Instruments)

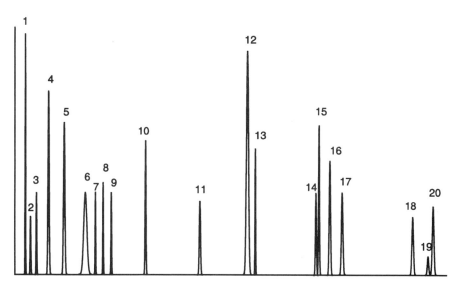

Figure 9.14 Results of a refinery gas analysis using MD system. Key: (1) hydrogen, (2) oxygen (3) nitrogen, (4) CO_2, (5) methane, (6) CO, (7) methane, (8) ethane, (9) ethylene, (10) propane, (11) cyclopropane, (12) isobutane, (13) n-butane, (14) tert-butene, (15) butene-1, (16) butene-1, (17) cis-butene-2, (18) isopentane, (19) n-pentane, (20) 1,3-butadiene. (Reproduced by permission of Varian Analytical Instruments)

results from two detectors, an FID and TCD, on a single chromatogram. Argon is employed as the carrier gas to improve the detectability and linearity for hydrogen. An interesting aspect of this instrument is the use of the **methanizer** which is a catalytic device situated between the column and detector to convert CO and CO_2 to methane quantitatively. These gases can be detected at the ppm level using the FID. Figure 9.14 shows a typical application result.

9.6 SUMMARY

Open tubular columns can be coupled in a variety of ways to perform separations which are impracticable with any single column regardless of its theoretical plate number. Column configurations can vary from simple series connected columns to more complex arrangements to allow heartcutting, backflushing or column selection for specific groups of sample components.

Multidimensional GC is obviously an important optional technique for these situations, but it is not widely employed in disciplines other than the petrochemical industry. Heartcutting techniques are in fact under-employed in practice as there are many analytical situations which could benefit.

For instance, an important requirement in the pharmaceutical industry is the determination of separate enantiomers in some products. By using MDGC the racemic fractions can be passed to a chirally selective column to separate the enantiomers.

Another possible application is general trace analysis where column A would be a high capacity column such as a thick film wide bore column. This is loaded with a relatively large sample volume and fractions containing the trace components passed to the second high resolution column in amounts sufficient for their measurement. Such situations are commonplace in environmental, food and beverage and clinical applications where the required components are present in small amounts in association with a complex matrix of interfering compounds.

REFERENCES

1. Deans, D. R. (1968) *Chromatographia*, **1**, 18.
2. Schomburg, G. and Weeke, F. (1972) in *Gas Chromatography 1972*, (ed. S. G. Perry), Applied Science Publishers, London, pp. 285–294.
3. Boer. H. (1972) in *Gas Chromatography 1972*, (ed. S. G. Perry), Applied Science Publishers, London, pp. 109–132.
4. Deans, D. R. (1981) *J. Chromatogr.*, **203**, 19028.

Index

Absolute retention data 20
Acids, derivatization of 230
Active sites 107, 111–113
  effect on retention indices 154
Activity coefficients 146
Aerosols 214
Air analyser 220
Alcohols 171
  derivatization of 230
Alkanes 132, 184, 202
  use of in retention index
    measurement 153
Aluminium oxide coated PLOT
    columns 162, 164–167, 286
  effect of gas velocity in 162
Amides 228
Amines 171, 174, 175
  derivatization of 230
Amino alcohols 143
Ammonia 171, 174, 175
Amphetamine enantiomers 231–232
Analysis speed 36, 134–135, 158
Analysers 57
Antidepressants 118
Atmospheric
  particulates 214, 242
  volatiles 226
Automatic flow control 62, 63
Autosamplers 9, 182, 209
  use in analysis 238
Average (mean) carrier gas velocity 12, 19, 256

and detector response 251

Back pressure regulator 59, 60, 61
Band spreading in space 205, 207
Baseline construction 104
  by perpendicular drop 104
  by tangential skim 104
Baseline instability 74, 263–265
Beverages 220, 288
Bonded phases 8, 117, 118
  and thickfilms 119
  on-column injection with 119
  splitless injection with 119
Boron trifluoride/methanol derivatization
    reagent 231, 232
Boyles law 16, 17
Butter triglycerides 120

Calibration procedure
  for gas analysis 236–237
Capillary column requirements 53
Carbohydrates, derivatization of 230, 233
Carbon coated PLOT columns 164, 165
Carboxylic acids 228
Carotenoids 3
Carrier gas 1, 8, 9
  and separation speed 38
  effect of choice 37
  inlet velocity 12
  optimum velocity 36, 122
  practical operating range 43

Carrier gas (*continued*)
  velocity of   12
  viscosity of   58, 123
Chemiluminescence detector (CLD)   97
  mechanism   98
Chiral phases   148–149, 231–232
Chromatogram   9
Chromatographic elution
    problem   252
Chromatography   3
  adsorption   2, 4, 217
  column   1
  column elution   2, 3
  displacement   4
  gas   1, 2
  gas–liquid partition   4, 125
  gas solid   1, 2, 125, 161
  high performance liquid (HPLC)   2, 217
  liquid   1, 2
  liquid–solid adsorption   2, 3
  paper   1
  partition   1–4
  thin layer   1
Clinical applications   6
Coating of PLOT columns   163
Column
  diameter   5, 11, 132
  features summary   158–159
  film thickness   5, 11
  holdup time   11, 19
  oven   5, 8, 9, 68
  oven properties   69
  overloading   125, 126
  permeability   17, 119, 121–123, 132, 140
  sample capacity   124–129, 179, 181, 189, 248
Column coupling   12, 58, 65, 67
  multiway   67
  reducing ferrules for   65, 66
  stainless steel ferrules for   65, 66
    using single ferrule
      connectors   67
  using soft ferrules   65, 66
  using taper-fit connections   65, 67
  using two-ferrule connectors   67

Column diameter
  choice of   132
  effects of   11, 39–42, 132
  and analysis speed   51, 134
Column efficiency, *see* Theoretical plate
    number
Column flow rate   54
Column installation   54
Column length
  and trace analysis   251
  choice of   130, 131
  effect in PTGC   258
  effects of   124, 129
Column switching   267
Columns
  aluminium clad   110
  capillary   2, 9
  fused silica open tubular   2, 8, 108, 109, 111, 158
  glass   107, 114
  narrow bore   133, 134
  open tubular   2, 5, 8, 10, 107–176
  packed   2, 122, 267, 268, 281
  packed capillary   2, 10, 121
  porous layer open tubular (PLOT)   2, 8, 10, 108, 125, 126, 160–176
  stainless steel   7, 107, 111, 119, 120, 260, 261
  support coated open tubular
    (SCOT)   160
  wall coated open tubular (WCOT)   2, 107–159
  wide bore   133, 135–140, 158, 209, 252, 281
Comparison of detector response
    characteristics   105
Compression fittings   58, 65, 67
Concurrent solvent evaporation   208
Conditioning of columns   108, 116, 117
Constant flow operation   63
Constant pressure operation   58, 63
Cooled needle split technique   202
Cryogenic focusing   270, 279
Cyanopropylsiloxane phase   118

Data system   9, 52, 61, 102, 103
Dead time, *see* Column, holdup time

INDEX

Detector
  contamination 263
  detectivity 75, 245
  dynamic range 75, 76
  linear range 75, 76, 105
  linearity 75, 76, 237
  make up flow 9, 55, 61, 76
  response 70, 247–248
  sensitivity 73, 87, 245, 247–248
Detectors 9, 69–99
  comparison of 105
  concentration sensitive 70–73, 88, 94, 245, 247
  flame 77
  selective 70
  specific 70
  universal 70, 94
Diameter effects 124, 127, 129
  in trace analysis 249
Diazomethane derivatization
  reagent 230
Diffusion coefficient in gas phase 30, 35, 41, 122
  at column exit 35, 122
  calculation by Gilliland equation 34
  values 37
Diffusion coefficient in liquid phase 30, 35, 122
Dioxins analysis by GCMS
Direct injection 140
Discrimination errors 182, 184, 185, 193, 202, 236
Distillation 211
  Dean and Stark 213
  fractional 212, 259
  micro 211, 212
  steam 211–213
  vacuum 212
Distribution (partition) coefficient 4, 10, 11, 145, 248
Drylab software for PTGC
  optimization 252
Dual detector operation 98
Dynamic coating of columns 107, 115

Effluents 220, 242

Electron capture (ECD) 8, 70, 87, 105, 232, 279
  mechanism 88
  operating features of 89
  radiactive sources for 90
  response towards halogenated compounds 91, 105, 201
Environmental monitoring 6, 99, 115, 144, 209, 211, 242, 288
EPA mixture 624
  optimized separation of by selectivity tuning 27
Exit flow rate 20
External standard method 236–237, 240
External standard method 238–241
Extra-column effects 53, 54, 55, 268
Fats 217
FFAP (free fatty acid phase) 115, 149
Film thickness 10, 107, 115, 124, 127, 129, 140–142, 158
  effects 39–42
  effect on trace analysis 249, 250
  effects in PTGC 258
Flame ionization detector (FID) 6, 8, 69, 77–83, 105, 232, 288
  design 77, 78
  effect of supply gases on response 80–83
  mechanism 79–80
Flame photometric detector (FPD) 86, 105
  design of 86
Flavour compounds 7
Flooded zone in splitless injection 198
Flow control system 8, 9, 194
Food samples 220, 288
Free fatty acids 115, 116
Freeze drying 212
Frontal analysis 4
Fronting peaks 263
Fused silica cold trap 215, 222, 279

Gas analysis 185–186, 236–237
  accuracy of 237
  using wide bore columns 236

Gas chromatography–mass spectrometry 99
  chemical ionization 99
  column coupling methods in 101, 102
  ion focusing in 100
  ionization processes in 99
  single ion monitoring in 100
  use of ion trap in 100
  use of quadrapole in 100
Gas compressibility 13, 16, 134, 135, 268
Gas supplies 57
  use of filters for 58
  gas cylinders for 58
  generators for 57–58
  purity of 58
Gas velocity 32
  average 122
  optimum 32
Gas viscosity 17
Gaussian shape of peaks 24, 26, 46
Giddings equation 35, 122
Glycols 228
  derivatization of 230
Golay theory 33–44, 235, 245
  effect of column diameter 33
  effect of gas velocity
  for wide bore columns 136
  mass transfer in PLOT columns 160, 161, 162, 163
  optimum values from 35–36

Halocarbons (EPA mixture 624) 272–276
Headspace analysis 177, 189, 212, 214, 219–226, 242
  automatic samplers for 222
Headspace sampling
  dynamic 221, 224–226
  purge and trap 221
  static 221–224
Henry's law 124, 222
Hormones, derivatization of 230
Hydrocarbons 167, 168, 171, 172
Hydrogen flame thermocouple detector 6

Immobilized phases 1, 117, 161
Impurities in 1,3-butadiene 165
Inert support 5, 10
Injection 8, 9, 68
  ash vaporization 68, 182
  auto 68
  bypass 68, 186
  direct 187, 188, 189, 209
  on column 68, 182, 187, 188, 203–209, 243, 252
  liners 194–196
  liners for wide bore 190
  procedure in analysis 235
  programmed temperature vaporization (PTV) 68, 187, 194, 202–203
  septum-less 186, 187
  split 68, 187, 188, 191–195, 209, 245
  split-splitless 68, 187, 188, 195–202, 204, 208, 243, 252
  time 192, 193
  vaporization 68, 181
Injector contamination 263
Instrument design 52
Integration errors and resolution 46–48, 50
Integration methods 103
Internal normalization 241–242
Internal standard 185, 238–242
  corrected 239
Ion pair extraction in sample preparation 212
Ion trap detector 100
Isothermal operation 8, 131, 235, 272, 281
IUPAC nomenclature 70

Katharometer, see Thermal conductivity detector
Ketones 171
Kinetic theory 20
Kozeny–Carman equation 121
Kuderna Danish concentrator 212, 213

Layer thickness (of PLOT columns) 10
Liquid chromatography–gas chromatography (LCGC) 212, 218–219

INDEX

applications of   219
MAOT (maximimum allowable operating temperature)   116, 117, 259
Martin compressibility correction factor   20
Mass flow regulator   59, 60
McReynolds constants   158
McReynolds probe compounds   156
Mean carrier gas velocity, *see* Average carrier gas velocity
Memory effects   180
Mercaptans   171
Methanizer   287–288
Microlitre syringe   177, 183, 193
  gas tight   186
Minimum detectability   75, 76, 105, 243, 245, 248, 250
  comparison of using different detectors   246
Mixed stationary phases   272
Molecular sieve PLOT columns   164, 165, 283, 285
Motor oil, simulated distillation of   260–261
Multidimensional GC   212, 267–288
  back-flushing in   267, 268, 276–278
  by pressure balancing system (Deans)   268, 277
  series connected column in   268
  switching valves for   268, 276, 280, 285, 287
  use for heart-cutting   267, 278–283
  with multicolumns   283–288

Naphthas by multidimensional GC   283
Narcotics, derivatization of   230
Natural gas analysis   283, 285–286
Natural products   3
Nitrogen compounds   83, 98
Nitrogen phosphorus detector (NPD)   70, 83, 105
  linearity of   85
  mechanism   84–85
Non-retained compound   12
Nucleotides, derivatization of   230

Optimization   235, 261–265

  of programmed temperature conditions   258–260, 272
Organochlorine compounds   196

PAH (polycyclic aromatic hydrocarbons)   120, 217
Partitions coefficient, *see* Distribution coefficients
Peak capacity   130
Peak measurement   102–105
Peak tailing   5–7, 55, 112–115, 179, 226, 228, 262, 263
Peak volume   54
Peak width   45
  at half height   28
  definition of   27
Permanent gases, separation of   95, 169, 170, 283
Pesticides   201, 252, 253–255
  in soil by SFE-GC   216–217
Pharmaceuticals   6, 118, 224
Phase ratio   11
Phenols, derivatization of   230
Phosphorus compounds   83
Photoionization detector (PID)   96, 105
  schematic principles   97
Plate theory   4, 20, 23, 246
Plunger in barrel syringe (PIB)   182
Plunger in needle syringe (PIN)   182
Poiseuille equation   121
Poisson distribution of chromatographic zone   2
Polar compounds   111, 228
  headspace sampling of   224
  tailing of   7
Polychlorinated biphenyls (PCBs)   144, 201
  in soil by SFE-GC   216–217
Polyethylene glycol (PEG)   6, 13, 115, 147, 149, 150
Polyimide   109, 111
Polysiloxane phase   120, 143, 149, 151, 260
Polywax 1000   110
Porapak (porous polymer phases)   172
Porous polymer coated PLOT columns   164, 283

Porous polymer phase   95, 170–174
Pre-pressure regulator   59, 60, 61
Pressure drop   5, 123
Pressure regulator   59, 60
Programme rate   69, 251
Programmed pressure operation   63, 64
Programmed temperature operation   58, 203–204, 235, 272
   and trace analysis   249, 250
   final temperature in   252
   initial temperature in   252
   programme rate in   252, 258
   theory   252–257
Proteins   3
Purge and trap techniques   139–140, 189, 212, 214, 220, 224–226, 242
   equipment for   225
   for fingerprinting of pillow down   227
   trace volatiles in rock by   226,
   aqueous samples by   226

Quantitative analysis   2, 73, 235–242

Raoult's law   2, 21
Referee methods   46
Refinery gases by multidimensional GC   283, 286–288
Relative retention data   20
Resolution   14, 44, 262, 265
   calculation of   48
   definition of   45
   effect of temperature on   151, 152
   equation   49, 129
   for quantitative analysis   235
Resolution map   254–255,   272, 273
Response factors   75, 242
   relative   75, 239, 240
   with external standards   236
Retention factor   11, 245, 249
   calculation of   20
   effect of boiling point on   152, 257–261
   effect of on column capacity   128
   effect of on mass transfer coefficients on   42
   effect of temperature on   150–152
Retention gap   67, 199, 206, 208, 209

   deactivation of   206
Retention index (Kovats)   152–158
   measurement of   154
Retention temperature   246, 248, 249, 255, 257
Retention theory   145–146
Retention time
   adjusted   11
   total   11
Retention volume   26
Rohrschneider–McReynolds
   characterization of phases   155, 158

Sample derivatization   226–233
   by acylation   228, 231–232
   by alkylation   228, 230–232
   by silylation   228, 232
Sample introduction, *see* Injection
Sample preparation   209, 211–233, 235
   errors of   236
Sample volume   236
Secondary cooling of injector   208
Selective electron capture
   sensitization   90
Selectivity tuning   270
Separation factor   13, 45
Separation number (Trennzahl)   131, 132
Septum   186, 194, 195
   flush (purge)   60, 61, 194, 197
Shampoo, determination of trace amounts of dioxane in   283, 284
Silylation reagents   229
Simulated distillation   257, 259–261
Size exclusion in sample preparation   212, 217
Soil samples   220, 242
Solid adsorbents for sample extraction   217
Solid phase extraction (SPE)   211, 212, 217–218
   adsorbents for   218
Solvent (focusing) effect   198–200
Solvent extraction   212–214
Soxhlet   212, 213
Split flow   9, 188
Split peaks   205

INDEX

Split ratio 60, 192, 193
  minimum 192
Squalane, use of in McReynolds
   constants 157–158
Stack gases 214
Standard deviation 24, 26
  of chromatographic zone 29
Standard methods of test 46
Static coating of columns 107, 115
Stationary phase 142–144
  characterization of 155–158
  coating 107, 115
  common 147–149
  definition of 1, 10
  effect of 129
  polarity of 146, 155–158
  selectivity of 21, 146
Steroids, derivatization of 230,
   243–244
Sub-ambient operation 69
Sublimation used in sample
   preparation 212
Sulphur odorants in natural gas 85
Supercritical fluid extraction coupled with
   capillary GC (SFE-GC) 212–217
Surface deactivation 108, 112–115
Surface roughening 114
Surface roughness factor 161
Syringe handling 183–185
  cold needle technique 183, 184
  filled needle technique 183
  handling (injection) time 192, 193
  hot needle technique 183, 184
  solvent flush technique 183, 184

Tenax adsorbents for sample
   extraction 225–226
Theoretical plate height 13, 23, 29
  of coupled columns 269
  minimum 32, 40
Theoretical plate number 4, 13, 20, 22,
   29, 178, 180, 209, 245
  calculation of 27, 28
  of coupled columns 269

Thermal conductivity detector (TCD) 4,
   5, 91–96, 288,
  carrier gases for 93, 94
  mechanism 91–93
  microTCD 93, 174
  nano TCD 93
  operating features 96
  response characteristics 94, 95, 105
Thermal desorption 189, 213, 220, 242
Thickfilm columns 40
Thinfilm columns 51
Tobacco smoke 214
Trace analysis 73, 209, 227, 230, 238,
   242–252
  by heart-cutting techniques 279, 288
  of impurities in propylene 244
  of nitrosamines 85
  of sulphur compounds 85, 86, 87, 98
  peak enhancement in 250–251
  recommendations 251–252
Triton–100 using on column
   injection 206
Troubleshooting 261–265
Trough ratio 47, 50
Trouton constant 152, 257
Tswett, origins of chromatography 3

Van Deemter theory 29–32
  eddy diffusion in 30
  longitudinal diffusion in 30, 31
  mass transfer terms 31, 32
  multipath term in 30
  non-equilibrium terms 32
Van der Waals forces 146
van't Hoff relationship 150
Void time 12
  see also Column, holdup time

Water 171, 172, 220
  analyser 220
Wine, determination of trace amounts of
   polyethylene glycol in 28

Zone dispersion 22, 42, 178